Neurosecretion

TERTIARY LEVEL BIOLOGY

A series covering selected areas of biology at advanced undergraduate level. While designed specifically for course options at this level within Universities and Polytechnics, the series will be of great value to specialists and research workers in other fields who require a knowledge of the essentials of a subject.

Titles in the series:

Biological Membranes	Harrison and Lunt
Water and Plants	Meidner and Sheriff
Comparative Immunobiology	Manning and Turner
Methods in Experimental Biology	Ralph
Experimentation in Biology	Ridgman
Visceral Muscle	Huddart and Hunt
An Introduction to Biological Rhythms	Saunders
Biology of Nematodes	Croll and Matthews
Biology of Ageing	Lamb
Biology of Reproduction	Hogarth
An Introduction to Marine Science	Meadows and Campbell
Biology of Fresh Waters	Maitland
An Introduction to Developmental Biology	Ede

TERTIARY LEVEL BIOLOGY

Neurosecretion

SIMON H. P. MADDRELL
JEAN J. NORDMANN

Department of Zoology
University of Cambridge

A HALSTED PRESS BOOK

John Wiley and Sons
New York – Toronto

Blackie & Son Limited
Bishopbriggs
Glasgow G64 2NZ

Furnival House
14–18 High Holborn
London WC1V 6BX

Published in the U.S.A. and Canada by
Halsted Press,
a Division of John Wiley and Sons Inc.,
New York

First published 1979

International Standard Book Number
0–470–26711–9

Library of Congress Catalog Card Number
79–63655

Printed in Great Britain by Thomson Litho Ltd East Kilbride Scotland

Preface

ANIMALS USE NERVE CELLS FOR THE CONTROL OF AND COMMUNICATION between different parts of the body. Everyone is aware of the way in which nerves transmit information at high speed by impulses travelling along their axons. This book is about the other, lesser known, but very important way in which nerve cells convey information over considerable distances. This they do by releasing substances, neurohormones, into the extracellular space. The phenomenon is termed neurosecretion and all animals use it to a greater or lesser extent. In fact, in evolution, it may well have predated nervous conduction as a means of communication within animals.

In this book we have attempted to avoid a wealth of detail in order to concentrate on the mechanisms and principles involved. We felt also that it was important to consider the generalizations one can make as to how neurosecretory systems work, and how they are designed so as to operate efficiently in conjunction with other physiological systems.

To a large extent we have concentrated on those areas which seem to us to be the most important and interesting. In adopting this approach we run the risk, of course, of neglecting areas which in the future may turn out to be of great importance. However, we feel that what may be lacking in completeness of cover is compensated by a deeper analysis of what we see as the most significant developments in the field.

In the last chapter we have picked out those areas where we feel more research is needed, in the hope that we may contribute something to the growth of what we think is a fascinating and most exciting area of animal biology.

The book is intended for undergraduates in their second and higher years reading physiology, anatomy, zoology or endocrinology. We think it will also be useful to those who do research in these areas, as a selective review of the subject. We have, in the body of the text, indicated our reference sources, which should make it easy for the reader to have direct access to what we think are the most authoritative and up-to-date pieces of work in the field.

Simon H. P. Maddrell
Jean J. Nordmann

ACKNOWLEDGMENTS

THE AUTHORS ARE INDEBTED TO A GREAT MANY PEOPLE WHO HELPED IN THE preparation of this book by allowing them to quote from their unpublished work, who read early drafts of some of the chapters, who discussed and argued various points with them, or who allowed them to use some of their original drawings or micrographs To all of them they gratefully acknowledge their debt. In addition they want particularly to mention Mrs. Margaret Clements and Mrs. Vanessa Rule who made light of the heavy typing and M. Roger Miguelez who is responsible for the better drawings in these pages.

Contents

CHAPTER ONE

THE PHENOMENON OF NEUROSECRETION

Introduction

In any metazoan animal it is of great importance that the activities of all the cells, tissues and organs are coordinated so that the animal functions efficiently. Cell-to-cell contacts play an important role in this. It has been found, for example, that the cells of some cancers make deficient contacts with their neighbours. For all except the smallest animals, however, the development of the longer-range coordinating mechanisms of the nervous and endocrine systems is just as important. The nervous system is particularly well adapted for rapid communication, while the endocrine system is more suited to exert control over relatively long-term bodily processes such as growth and development, or the control of the composition of the internal environment. However, the use of the nervous system is not confined to rapid responses.

The idea that many nerve cells synthesize and release long-lived potent hormones may come as a shock to anyone who thinks of nerve cells solely as high-speed devices involved in rapid communication. Except where they make electrical synapses, however, all nerve cells convey information to other cells by releasing special chemical substances. Where nerve cells release their products at some distance from their target cells, the substances they release must have a longer life so that they can build up to an effective concentration. Where the distance between nerve cell and target is so large that release occurs into much or all of the extracellular space of the animal, the substance released qualifies as a true hormone. It is nerve cells whose primary function is the synthesis and release of hormones (or other substances which act at a greater distance than at a synapse) that are described as *neurosecretory*.

If nerve cells can communicate very rapidly with other cells by conducting impulses down axons, why do some indulge in the slower mechanism of secretion of hormones? To answer this, one has to reflect on the requirements of the communication between nerve cells and target cells.

The message which is transmitted may be private to one, or general to a great many target cells; it may need to be very rapidly delivered, or it may be one where this is not critical; and, finally, it may be important for the message to be very short or to be a long sustained one. The nervous system is modified in distinct ways to provide these attributes of specificity, speed of delivery, and length of message. The characterizing features of a nerve cell are that

(1) it can pass information along itself electrically at high speeds in the form of nerve impulses,
(2) it can release a chemical effector substance on the arrival of the impulses at its endings.

By varying the extent to which these two activities are emphasized, the nervous system can deliver information which varies in the three qualities mentioned. Figure 1.1 illustrates the ways in which various types of

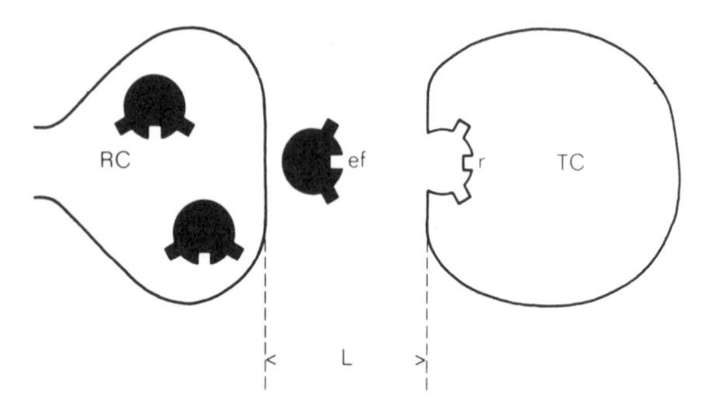

Figure 1.1 To show how a release cell (RC) can affect a target cell (TC) in a variety of ways. The speed of the process depends on the distance (L) between the two cells. The specificity of the effect depends on the interactions of the effector compound (ef) and the receptor site (r) on the target cell. An effect of short duration is produced when release ceases promptly and the effector compound is rapidly diluted and/or destroyed and excreted. For a more prolonged effect, release is maintained and/or the effector is stable and not rapidly removed from the extracellular fluid.

messages are provided for, and makes the additional point that the target cells themselves and other systems (such as destructive enzymes and excretion) are also involved in defining the type of message that is delivered. The choice between alternative ways of achieving a result is usually determined by the speed of delivery required. A widespread effect, for example, can very rapidly be achieved by the simultaneous release of effector materials close to a large number of cells or, if the speed of delivery is not important, a more leisurely release from relatively fewer cells into the general circulation is adequate.

Historically, the term *neurosecretion* has been reserved for those cases in which there is clear cytological or histochemical evidence in nervous tissue of synthesis of material which, it is presumed, the nerves later release. However, as emphasized above, very nearly all nerve cells synthesize and release specific substances. From the use of the electron microscope and from our increasing knowledge of how and where in nerve cells the synthesis of such specific substances goes on, it has been possible to find the sites of this synthesis and storage in all the nerve cells that have been investigated. In other words, practically all nerve cells can be thought of as neurosecretory. Such a view has in fact been held for some time.[1]

While it is true that all nerve cells behave fundamentally in a similar fashion, there are, nonetheless, at the extremes of the range of capabilities of the nervous system, modes of operation which are capable of separate treatment. All of us are aware of the ways in which the nervous system rapidly delivers highly specific short messages. The main object of this book is to describe the phenomena at the other end of the spectrum, i.e. the ways in which the nervous system produces a relatively general effect for more sustained periods. Since this involves largely only those nerve cells that by the older techniques could be set apart as neurosecretory neurones, this is our excuse for continuing to refer to these activities under the term *neurosecretion*. In the chapters that follow, then, we shall be concerned with those nerve cells which release their specific products at some distance from their target cells. The products released by these cells in most cases are true hormones, though some of them have less than the wide circulation throughout the organism which the term *hormone* implies. Unfortunately, the number of cases in which hormones are known to be released is still rather small, so that to some extent we have again relied on the fact that there is a reasonably good correlation between the cytological appearance of a nerve cell and whether or not it releases a hormone or other compound with a relatively widespread effect. In general, neurosecretory neurones can be recognized by a number of specific histochemical techniques[2] (such as staining with chrome-haematoxylin-phloxine or with paraldehyde-fuchsin) and, at the ultrastructural level, by the presence within the cell body of large numbers of so-called "elementary neurosecretory granules" which are membrane-bound vesicles usually about 100–300 nm in diameter with contents which characteristically are electron-dense. An electron micrograph of a typical neurosecretory cell is shown in figure 1.2.

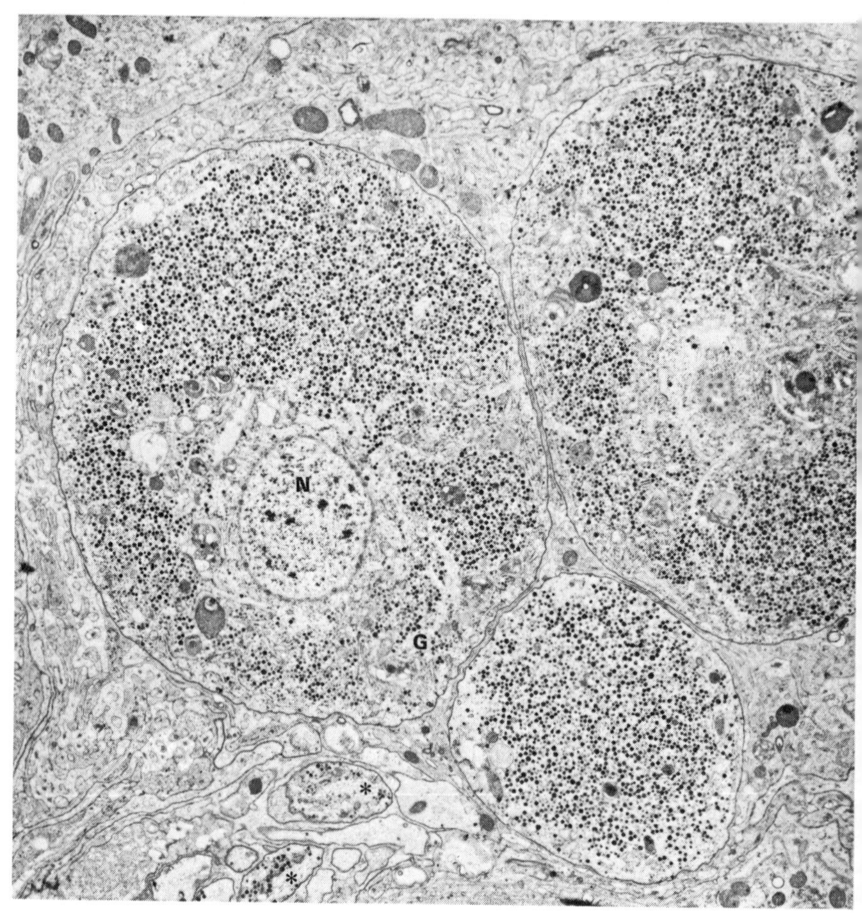

Figure 1.2 Neurosecretory cells in the central nervous system of the insect *Rhodnius prolixus*. Note the cell bodies with their high content of electron-dense neurosecretory granules. Two axons from the cells can be seen in the lower part of the micrograph (*). G-Golgi complex; N-nucleus. (Micrograph courtesy of C. G. H. Steel & G. P. Morris. × 8375)

The general features of neurosecretory cells and neurosecretory systems

Neurosecretory cells, as with other specialized cells, are much modified in form, structure, and behaviour to suit them to their function. How do they differ from "ordinary" neurones?

In general, neurosecretory cells make no close efferent synaptic junctions or contacts with other cells. Their endings are not very different from those of other nerve cells and, as we shall see, the events that go on there are in many respects similar to those at synapses. However, the target cells are usually not those that are closest to the endings. The "synaptic cleft" of a neurosecretory cell has become much bigger than in a conventional synapse, often to the extent that it involves the whole of the extracellular space of the animal.

A consequence of the large space between neurosecretory cell endings and their targets is that neurosecretory cells have to produce relatively much larger amounts of *transmitter substance* than do conventional neurones. Also, because the released products are not structurally confined to the region of the axon endings, they cannot be reabsorbed and re-used as happens, for example, with choline and acetate at cholinergic synapses. Instead, neurosecretory cells manufacture large amounts of neurosecretory material in their cell bodies. The structural appearance of the cell bodies is dominated by this. They contain numerous Golgi systems spawning great numbers of secretion-packed granules. It is the content of these granules which gives neurosecretory cells their characteristic staining properties. Because the diameters of the granules are close to the wavelength of the blue end of the visible spectrum, they scatter such light and give the cells their characteristic opalescent blue-white appearance.[3] This property is useful in recognizing neurosecretory cells in the living state so that they can, for example, be isolated for assay of their contents or be impaled with microelectrodes.

The neurosecretory granules which are produced in the cell body have to be transported to their sites of release at the axon endings. This involves the development of a system of axonal transport which will ensure the delivery of the granules at an appropriate rate and density to each of the release sites.

A further consequence of the large "synaptic cleft" of neurosecretory cells is that release at one or a few axon endings would be a very slow way of achieving effective concentrations of the particular product in circulation. It appears not to have been possible to evolve a more rapid way of releasing material at a single ending, so the evolutionary solution has been to increase enormously the number of axon endings from which release occurs.

Nerve cells, like man-made control systems, function best in a regulated environment. In many animals, a specially developed system provides them with this.[4] It usually involves some form of barrier between the general extracellular space and the immediate environment of the nerve cells. This *blood-brain barrier*, as it is often called, is obviously an embarrassment to neurosecretory cells whose function is to release material into the general

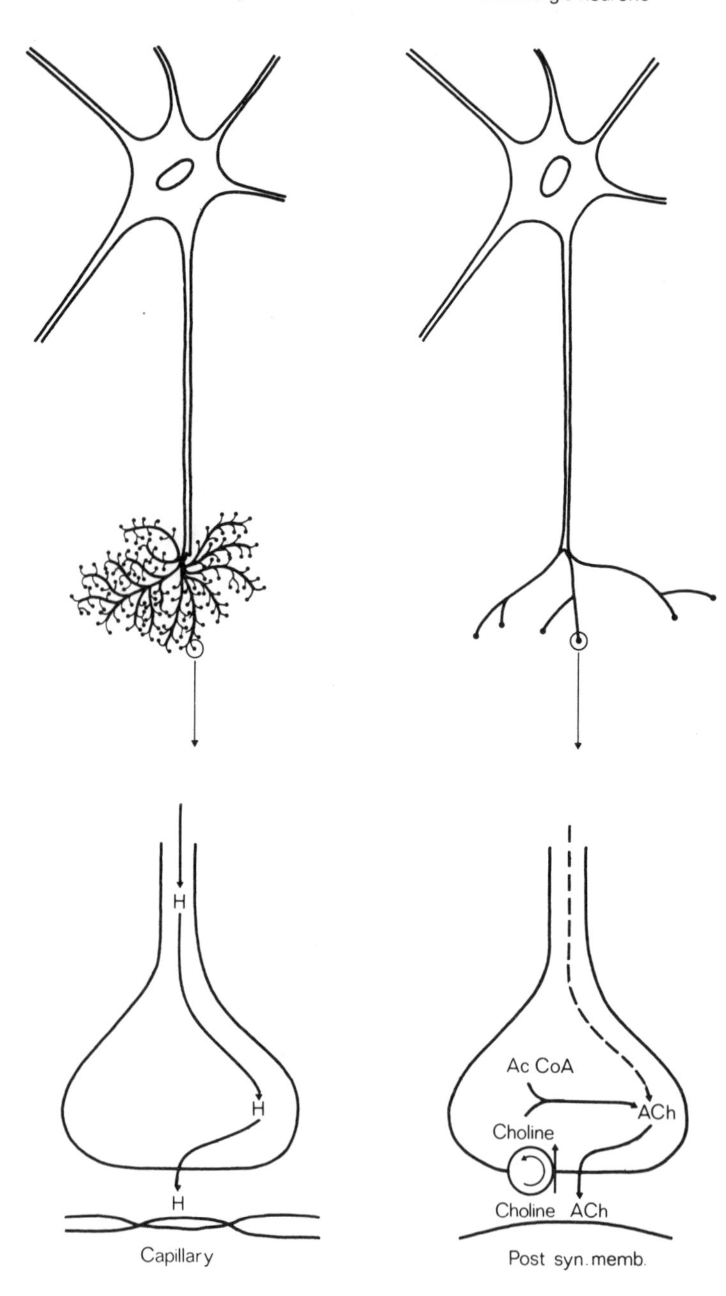
Typical neurosecretory cell
Cholinergic neurone
H
H
H
Capillary
Ac CoA
ACh
Choline
Choline
ACh
Post syn. memb.

body space. Special so-called *neurohaemal organs* occur in many animals where the blood-brain barrier is breached, and the neurosecretory axons branch and proliferate in contact with the general extracellular space.

Many neurosecretory products are only needed at particular stages or times within the animal's lifetime. In such cases of discontinuous release, neurosecretory granules are stored in the axon endings and may back up into the axons and cell bodies as well. There are obvious feedback problems in controlling such storage and regulating the turnover and manufacture of replacement granules.

The differences between neurosecretory cells and other nerve cells are summarized in figure 1.3.

We have attempted in this book to cover those areas which we think are the most interesting and important, and in which the most recent work has been done. In chapter 2 we describe the synthesis of neurohormones in the cell bodies of neurosecretory cells, the packing of hormone into membrane-bound granules, and the transport of these granules to the axon endings from which they are released. We also describe those modifications of neurosecretory cells which allow them to release enough hormone to achieve an effect. Chapter 3 covers the electrical events in neurosecretory cells that give rise to hormone release. Chapter 4 is concerned with the mechanisms whereby hormone is released from intracellular granules into the extracellular fluid. In chapter 5 the approach becomes a comparative one and we survey the different ways in which the neurosecretory system is arranged and used in different animals. Chapter 6 looks at the range of processes controlled by hormones in animals, how modulation of hormone secretion can have different effects, and examines the known structures of different neurohormones. Chapter 7 describes hormone-receptor site interactions and the mechanisms whereby these interactions lead to responses in the target cells. Chapter 8 deals with the problem of controlling the concentration of neurohormones in circulation, covering especially feedback effects. Finally, chapter 9 summarizes our main conclusions and emphasizes those areas where more research is needed.

Figure 1.3 To show the main differences between a typical neurosecretory cell (left) and a non-neurosecretory nerve cell (right)—in this case a cholinergic neurone. The neurosecretory cell has a very large number of nerve endings, each releasing into circulation hormone (H) which is synthesized in the cell body and transported along the axon to the endings. The cholinergic neurone has many fewer endings which release transmitter (acetylcholine) into the very limited space of the synapse; the transmitter is assembled in the nerve endings themselves so that axonal transport is not so important.

REFERENCES

1. De Robertis, (1964), *Histophysiology of Synapses and Neurosecretion*, Oxford, Pergamon Press.
2. Prentø. P. (1972) "Histochemistry of neurosecretion in the pars intercerebralis-corpus cardiacum system of the desert locust," *Gen. Comp. Endocr.* **118,** 482–500.
3. Thomsen, E. (1952) "Functional significance of the neurosecretory brain cells and the corpus cardiacum in the female blowfly, *Calliphora erythrocephala* Meig.," *J. exp. Biol.* **29,** 137–172.
4. Abbott, N. J. & Treherne, J. E. (1977) "Homeostasis of the brain microenvironment: a comparative account." In *Transport of Ions and Water in Animals* (eds: B. L. Gupta, R. B. Moreton, J. L. Oschmann & B. J. Wall), pages 481–510, London, Academic Press.

General references for background reading

(a) on nervous systems & neurosecretion

Berlind, A. (1977) "Cellular dynamics on invertebrate neurosecretory systems," *Int. Rev. Cytol.* **49,** 171–251, (pages 172–181 are relevant).

Bullock, T. H. & Horridge, G. A. (1965), *Structure and Function in the Nervous Systems of Invertebrates*, Vol. 1, San Francisco & London, W. H. Freeman. (pages 1–10 are relevant to this chapter).

Mason, C. A. & Bern, H. A. (1977) "Cellular biology of the neurosecretory neuron" In *Handbook of Physiology*, Section 1. *The nervous system*, Bethesda, Amer. Physiol. Soc., 651–689.

Morris, J. F., Nordmann, J. J. and Dyball, R. E. J. (1978) "Structure-function correlation in mammalian neurosecretion", *Int. Rev. Exptl. Pathol.* **18,** 1–95 (pages 2–70 are relevant).

Normann, T. C. (1976) "Neurosecretion by exocytosis," *Int. Rev. Cytol.* **46,** 1–77, (pages 1–20 are relevant).

(b) on animal structure and physiology

Eckert, R. & Randall, D. J. (1978). *Animal physiology* W. H. Freeman, San Francisco. (An excellent recent text for background physiological and cytological material.)

CHAPTER TWO

NEUROHORMONES PRIOR TO RELEASE

IN THIS CHAPTER WE SHALL DEAL WITH THE SEQUENCE OF EVENTS COVERING the synthesis of hormones in neurosecretory cell bodies, the packing of the hormone into granules, and the transport of these granules from the cell bodies to the nerve endings where they are accumulated prior to their release. We shall also follow the fate of granules which are not liberated. Finally, we shall see how neurosecretory cells are adapted so as to release enough hormone to achieve an effective concentration in the extracellular space.

Synthesis of hormone

As might be expected, ultrastructural studies show that the cell bodies of neurosecretory cells are rather similar to other active secretory cells, i.e. they have well-defined Golgi complexes, there is abundant rough endoplasmic reticulm and there are membrane-bound granules. Presumably, as with other secreted proteins, the hormone precursor is actually assembled from amino acids in the rough endoplasmic reticulum. From there it is transferred to the Golgi complex in small transporting vesicles pinched off the rough endoplasmic reticulm. Next, these vesicles are thought to fuse with the membranes of the Golgi complex so that their content of hormone precursor enters the Golgi saccules (figure 2.1). The Golgi saccules are arranged in stacks packed to form a shallow cup. The saccules nearest the concave side of the cup gradually fill up with secretory material so that they appear more and more dense. Finally, membrane-bound granules bud off from the saccules and move away into the cell cytoplasm (figure 2.1).

We have so far referred to the material found in the neurosecretory granule as *hormone precursor*. There is a good deal of evidence that by the time the granules arrive at the nerve terminals not only do they contain hormone in an active state but, in addition, one or more hormonally inactive proteins[1] and ions such as Ca^{++} and Mg^{++}. Because of the relatively large molecular size of these other compounds they must,

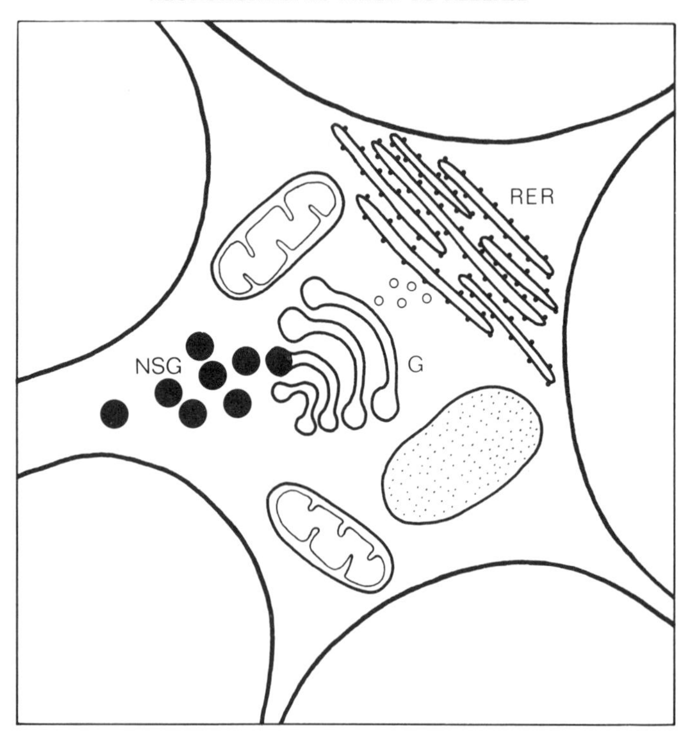

Figure 2.1. Synthesis and storage of neurohormones. Hormones are synthesized as a precursor on the rough endoplasmic reticulum (RER) and then pass through the Golgi complex (G) into condensing vacuoles. The hormones leave the Golgi complex packed in membrane-bound vesicles, the neurosecretory granules (NSG).

presumably, have been packaged together with the hormone in the cell bodies. In the case of the neurohypophysis, the best-known neurohaemal organ of vertebrates, the proteins present in the granules with the hormones, oxytocin and vasopressin, are called *neurophysins*. The neurophysins are rich in cysteine, a peculiarity which can be turned to advantage as they can be labelled with ^{35}S. It also accounts for the specific staining reactions of the cells. The neurophysins and the hormones found with them have a ready affinity for each other, so it has been very reasonably suggested that neurophysins by binding with the hormones might serve to prevent their diffusive loss from the granules in the time between their synthesis and liberation. Another view has been that neurophysins are found in neurosecretory granules, because they are products of the synthetic process by which the hormones are made. It is conceivable that oxytocin and neurophysin are, for example, first synthesized as long-chain peptides which are enclosed in granules in their respective cell bodies and then, at some later stage, the peptide chain is cleaved, perhaps by an enzyme, to give hormone plus neurophysin.[2,3] These two possibilities are not mutually exclusive, of course; the synthesis from a common precursor of a hormone and the protein which binds it would have the obvious advantage that the binding protein would appear at just the right time in amounts directly related to the amount of hormone needing to be bound.

If oxytocin and vasopressin and the neurophysins do arise from the same precursors after granule formation, then one might expect to find a specific neurophysin associated with each hormone, and also to find that the relative quantities and rates of synthesis of these two types of compounds are closely related. The following evidence provides support for this idea.

After rats were given injections of radiolabelled amino acids, either into the ventricles or the hypothalamic areas of the brain, radioactive vasopressin, radioactive neurophysin and radioactive neurosecretory granules all appeared in the neurohypophysis with very similar time courses.[4] Furthermore, the radioactivity associated with hormones and with neurophysins was then lost from the neural lobe at rates which were not significantly different.

When ^{35}S-cysteine was injected adjacent to the surpraoptic nucleus, it was rapidly converted into a single protein of molecular weight about 20 000.[5] With the passage of time and as the material passed down the pituitary stalk into the neurohypophysis, the major labelled protein changed (figure 2.2). By the time (2 h) the material reached the posterior pituitary, most of it was present as a protein of molecular weight about 12 000. By 24 h this was the only labelled protein present. It was possible to show that both these proteins were immunoreactive with antibodies to rat neurophysin. This strongly suggests that the 20 000-dalton substance is the precursor and the 12 000-dalton protein is the neurophysin itself. In addition to the appearance in the pituitary of labelled oxytocin and vasopressin, several other labelled peptides are found. All these are stored in and released by the posterior pituitary. What is the function of these other peptides is an intriguing question for the future.

An elegant piece of evidence that hormones and neurophysins have a common precursor is that inhibition of neurophysin synthesis by using analogues of amino acids (e.g. methyleucine and γ-methyl-methionine) that occur in neurophysin, but not in vasopressin, also inhibited the synthesis of vasopressin.[6]

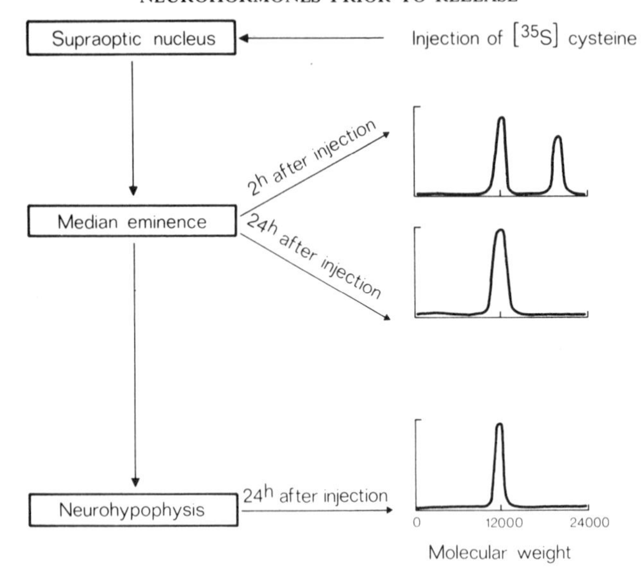

Figure 2.2 Evidence for a precursor in the biosynthesis of neurophysin. ^{35}S-cysteine, a protein marker, was injected close to the supraoptic nucleus of rats. Two hours later a large molecular weight molecule (20 000 daltons) was found in the median eminence. 24 h after injection, the 20 000 dalton molecule had disappeared and all the radioactivity was found in a single peak corresponding to that of the neurophysin. Note that no precursor was found in the neurohypophysis which suggests that maturation of the precursor had occurred during axonal transport (after Gainer *et al.*, ref. 5).

All these results support the idea that hormone and neurophysin are first synthesized together as different parts of a larger protein which is cleaved during transport in granules down the neurosecretory axons.

Work on isolated neurophysins from ruminants, from the pig and from the rat, all shows that there are specific neurophysins for each hormone. One line of evidence is that in mutant rats which suffer from hypothalamic diabetes insipidus, and which synthesize no vasopressin,[7] no vasopressin-neurophysin can be found.[8] In rats which are heterozygous for the condition there is an intermediate level of vasopressin-specific neurophysin.

Further evidence comes from using immunocytochemical techniques. Antibodies to proteins isolated and separated from the posterior pituitary react with two different populations of cells in the hypothalamus (figure 2.3).[40, 41] Presumably these cells are the ones which synthesize vasopressin and oxytocin.

Active hormone might be produced from its precursor at any stage after

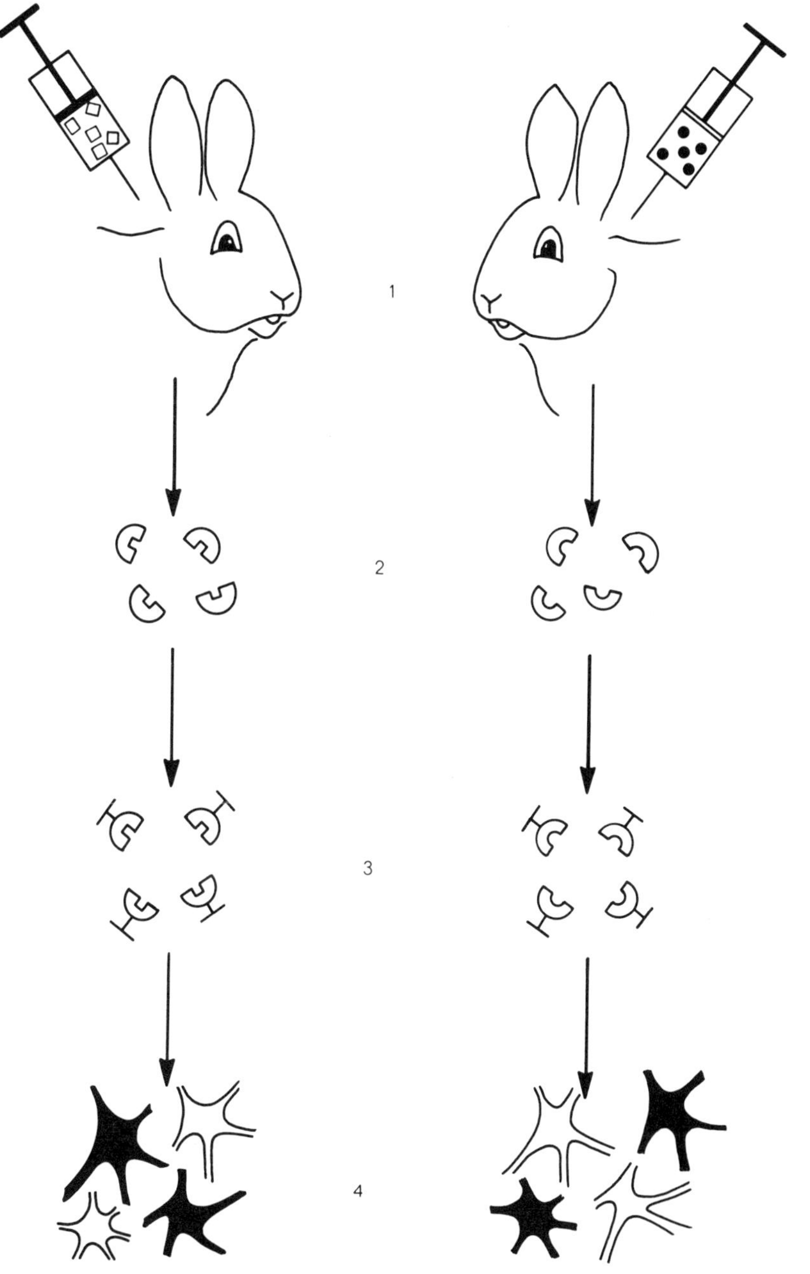

Figure 2.3 The use of neurophysin-specific antibodies to reveal oxytocin (left) and vasopressin (right) secretory cells in the hypothalamus.
1, injection of separated and purified neurophysins;
2, production of antibodies;
3, isolation of antibodies which are coupled to a marker such as horseradish peroxidase;
4, antibody-antigen reaction in hypothalamic neurosecretory cells.

the formation of the neurosecretory granules and before their arrival at the endings. Since much of the "maturation" of hormone precursors occurs during axonal transport, it seems likely that newly made granules do not spend long in the neurosecretory cell bodies. The evidence for this maturation of the prohormones in the hypothalamo-neurohypophysial system is that in the dog and the rat the ratio of active oxytocin to active vasopressin is much lower in the hypothalamus than in the neural lobe. This, presumably, reflects a faster rate of maturation of pro-vasopressin than of pro-oxytocin. In support of this idea is the finding that if granules are prevented from leaving the hypothalamus by removal of the hypophysis, then the level of active oxytocin in the hypothalamus rises more than does the level of vasopressin. Another relevant finding is that neurosecretory granules in the neural lobe respond differently to fixation compared with granules in the cell bodies.[9] At pH 8·0, aldehyde fixation gives very poor fixation of granules in the neural lobe, but the granules in the cell body are fixed in a much more intact state. Furthermore, the same fixation procedure gives better results for granules in the paraventricular nucleus of the hypothalamus, where more oxytocin than vasopressin is synthesized, than in the supraoptic nucleus which synthesizes more vasopressin than oxytocin. Again it looks as if pro-vasopressin matures more rapidly than does pro-oxytocin, though it should be pointed out that the same results would be obtained if axonal transport rates were slower in vasopressin-synthesizing neurosecretory cells than in those that produce oxytocin.

Work on other neurosecretory systems, such as those of insects, has produced a series of findings which are compatible with the way in which the vertebrate hypothalamus/neurohypophysis system is thought to operate. For example, histochemical techniques have shown that neurosecretory cells in the brain of locusts contain cysteine-rich proteins.[10] Recent determinations of the structure of such insect hormones as a cardiac accelerator, a hormone producing diapause (a seasonal dormant state) and a hormone causing diuresis, have shown that they are peptides of molecular weight ranging from 1000–3000;[11] oxytocin and vasopressin are peptides of molecular weight about 1000. Extracts from the brain or other parts of the central nervous system which contain the neurosecretory cell bodies but not the axon terminals (which, as we shall see in chapter 5, p. 79, lie in neurohaemal organs close to, but separate from, the main ganglia of the central nervous system) contain active substances of much higher molecular weight.[12] Possibly these are complexes of hormones with proteins similar to vertebrate neurophysins.

As in vertebrates, it has been possible in cockroaches to prepare antisera of protein extracts from neurosecretory material.[13] These antisera have been used to label individual neurosecretory cell bodies and their axon terminals. This technique is turning out to be very useful in localizing the cells responsible for synthesizing particular hormones, as well as the neurohaemal sites where their release occurs.

There is some evidence which suggests that insect neurosecretory granules "mature" during axonal transport, as do those of vertebrates. First, in some insects it is found that neurosecretory granules in the cell bodies differ in their histochemical reactions from those found in the axon endings, suggesting that the properties of the granule contents alter during transport.[14] Second, the hyperglycaemic hormone of locusts, which raises blood sugar levels, can be extracted in an active form from the corpora cardiaca containing the axon terminals but not from the brain where lie the cell bodies which synthesize the hormone.[15] In the bloodsucking insect *Rhodnius*, however, active diuretic hormone can be extracted either from the appropriate isolated single-cell bodies or from the nerve terminals.[16,17] In this case it might be that the prohormone has some activity, or it could be that newly synthesized neurosecretory granules are slow to leave the cell bodies, so that more maturation occurs before transport. The system is, in any case, rather unusual as the diuretic hormone is only released at a high rate when the insect takes a blood meal, and this may only occur six or seven times in its life time, with intervals of up to six months or even a year between meals. Finally, it is worth mentioning experiments on locusts where although injected radiolabelled cysteine was extensively incorporated in neurosecretory material of the corpora cardiaca (p. 17); little radioactivity was found to be associated with active hormone, most of it being incorporated instead in the neurophysin-like proteins associated with the hormones.[15] Presumably the hormones contain little or no cysteine, unlike the vertebrate pituitary neurohormones (see pp. 12, 138).

Axonal transport

Because neurohormones are synthesized in the cell body and released from the nerve terminals, some mechanism is needed to transport the hormone-containing granules along the axons away from the cell body. This transport is just one aspect of the necessity for nerve cells to have efficient internal transport systems because of their great length. Transport of granules away from the cell body can be demonstrated in experiments where ligatures are put on a nerve tract containing neurosecretory axons.

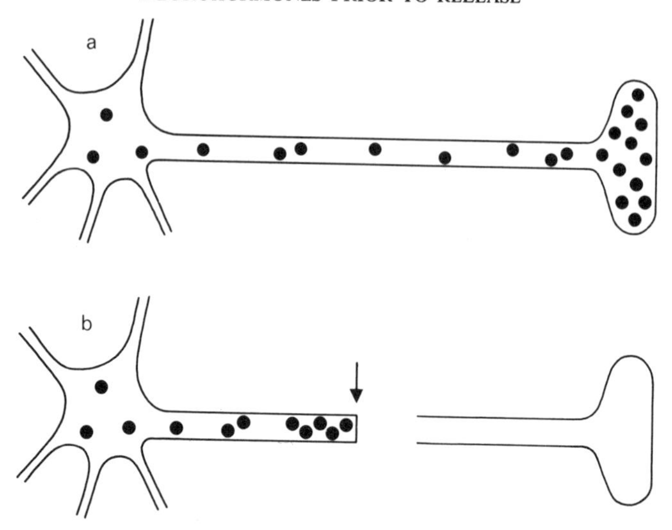

Figure 2.4 (a) Axonal transport in a neurosecretory cell. (b) After section or ligature of the axon, there is an accumulation of neurosecretory products in the axon.

When, after a few hours, the tissue is fixed, sectioned and stained, large amounts of neurosecretory material can be seen to have accumulated on the cell body side of the ligature[18] (figure 2.4). Such experiments have been performed, not only in vertebrates but also in several insects, among them the cockroach, for example.[19] Very similar effects are produced when nerve trunks are damaged locally and there is some lesion of the neurosecretory axons; again, neurosecretory material piles up in the axons just short of the damaged regions.

Because of its importance, the flow of materials within nerves has been the subject of a great deal of work within the last few years.[20, 21] It turns out that transport along axons is usually of two kinds; it can proceed slowly (at rates of the order of 1 mm day^{-1}) or at much higher rates (rates of 100–700 mm day^{-1} are commonly found). The distinction is beginning to look rather an arbitrary one, as several cases are now known where transport occurs at intermediate rates. Since neurohormones are only useful when they reach the nerve endings, the points in the cell furthest from the site of synthesis, it is perhaps not surprising that the hormone-rich neurosecretory granules are moved by the fast transport system. In the goldfish, neurosecretory granules were found to move from cell bodies in the preoptic nuclei down to the neurohypophysis at rates of up to 2 mm min^{-1} during electrical stimulation of hormone release from the nerve terminals.[22] This is the highest value so far recorded and, unusually in this case, it appeared that stimulation had accelerated the rate of flow; in most other systems stimulation seems to have little or no such acceleratory effect. Among work on insects, mention may be made of experiments on adult locusts where injected ^{35}S-cysteine was taken up into the brain neurosecretory cells.[23] Radioactivity appeared in the neurosecretory axon terminals in the corpora cardiaca with a time-course similar to that of its appearance in the cell bodies, but with a lag of about an hour. Presumably this lag was a measure of the rate of passage of material along the axons; from the distance travelled the rate of movement was about 40 mm day^{-1}. In the bloodsucking bug *Rhodnius*,[24] hormone release *in vivo* produced a series of histochemical changes compatible with an axonal flow rate of 120 mm day^{-1}.

Microtubules and axonal transport

Inhibitors of mitosis such as colchicine and vinblastine have disruptive effects on the organization of microtubules. Since these agents also block fast axonal transport systems, it has reasonably been suggested that the microtubules, which run longitudinally along the whole length of axons, might play an important role in the transport. Indeed one can often see granules in longitudinally aligned arrays in nerves running parallel to the microtubules (figure 2.5). In recent experiments on the splenic nerve of the ox, chains of neurosecretory vesicles, which can be seen in electronmicrographs of the nerves, were found to be robust enough to survive homogenization and centrifugal fractionation procedures. Linked vesicles have also been seen in nerves from lampreys and molluscs. Examination of the links holding the vesicles in chains showed that the vesicle membranes are joined by short (*c*. 2 nm) extensions of their outer membranes. By no means all neurosecretory granules in axons are joined to one another, but the fact that some are has implications for suggestions as to the mechanism of transport.

Just how microtubules and the neurosecretory granules might interact in such a way as to drive the granules along the axon is, so far, not at all clear.[21] The facts that tubulin, which in the polymerized form constitutes microtubules, can undergo phosphorylation and that microtubules *in vivo* can almost certainly bind calcium ions are reminiscent of the involvement of calcium and ATP in a good many other movement-producing systems (such as muscle contraction, ciliary beat, and in the contraction of the stalk of the sessile vorticellid protozoans). It has even been found that proteins

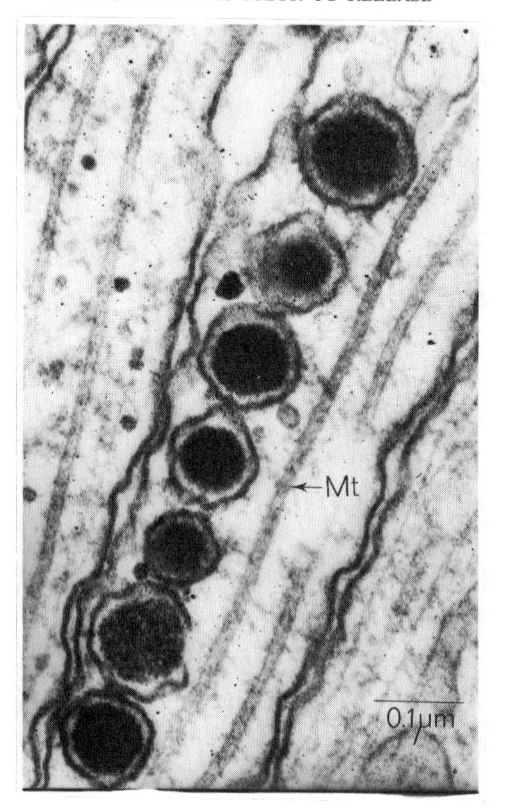

Figure 2.5 Neurosecretory granules in a neurosecretory axon (in this case of the mollusc *Anodonta*). The granules occur in a chain close to a microtubule (Mt). (Micrograph courtesy of E. Chipchase.)

similar to the actin and myosin of muscle can be isolated from nervous tissue, and that together the proteins act as a calcium-stimulated ATP-ase. These findings have led to speculation involving a series of arms projecting from the microtubules able to form cross-bridges with neurosecretory granules to propel them along the length of the microtubules (figure 2.6).

These speculations are intriguing, but clearly more evidence is required to determine the exact relationship between microtubules and neurosecretory granules.

The reader may have been surprised to read (p. 17) that the fast axonal transport which is responsible for carrying neurosecretory granules to the sites of release seems to occur no faster either during or after stimulation of hormone release. A large part of the explanation of this apparent paradox is that only a fraction of the neurosecretory granules synthesized are ever released; synthesis and axonal transport in most systems occurs rather faster than needed to meet normal release requirements, so that high rates of release for short periods can be met without any immediate acceleration of function in the supporting systems. These findings raise several new problems. What happens, for example, to granules which are not released; and are there any effects on, say, the rate of hormone synthesis after prolonged stimulation of release?

Fate of transported granules

Obviously the purpose of axonal transport is to carry neurosecretory material to its site of release, and in the next chapter we shall see how this

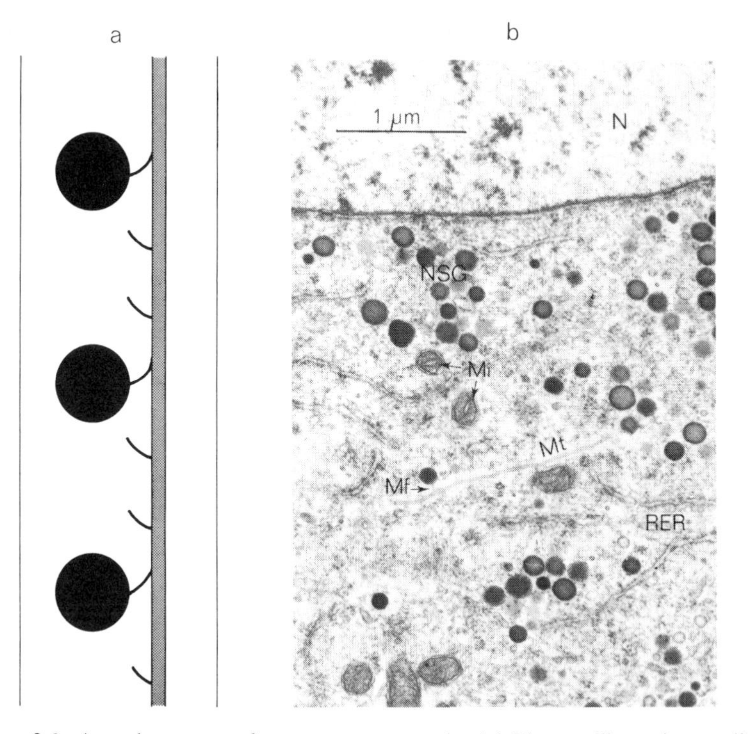

Figure 2.6 Axonal transport of neurosecretory granules. (*a*) Diagram illustrating possible transport by means of a propelling mechanism involving microtubules and side-arms. (*b*) A neurosecretory granule attached by a microfilament to a microtubule in *Helix aspersa*. Mi, mitochondria; Mf, microfilament; Mt, microtubule; N, nucleus; NSG, neurosecretory granules; RER, rough endoplasmic reticulum. (Micrograph courtesy of G. Nicaise.)

liberation is brought about. It is becoming clear that during secretion it is the contents of the most recently synthesized granules which are released first, followed, if necessary, by those of a slightly greater age. The evidence for this comes from experiments on rats, in which ^{35}S-labelled cysteine was injected into the third ventricle of the brain. This compound is incorporated into those proteins which are being synthesized in this region, among them those packed into neurosecretory granules in the hypothalamus. On stimulation of hormone release from the neurohypophysis, it was found that the specific radioactivity of the released material was *higher* than that of material remaining in the gland.[6]

These experiments also allowed the subsequent movements of the freshly labelled granules to be followed autoradiographically or biochemically. It was found (figure 2.7) that neurosecretory granules which arrive in the

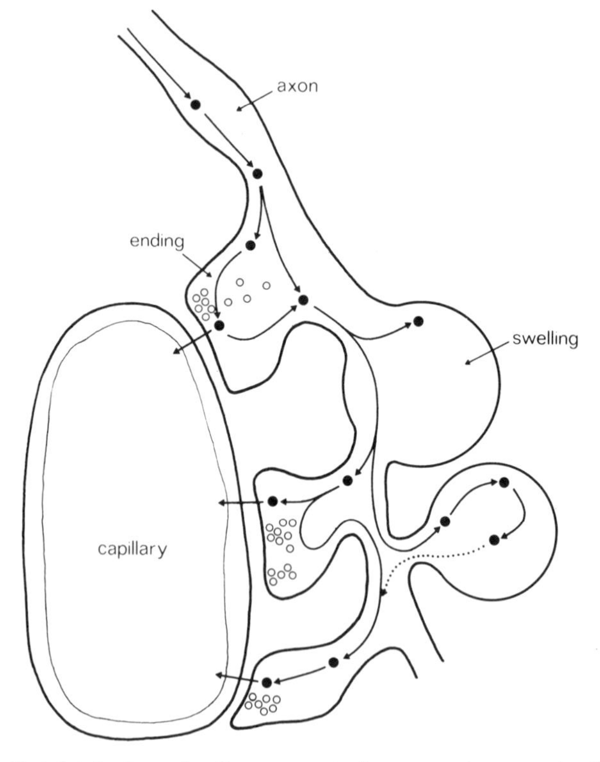

Figure 2.7 Postulated scheme for the movement of neurosecretory granules through the compartments of the hypothalamo-neurohypophysial neurone within the posterior pituitary (after Heap *et al.*, ref. 25).

axon terminals and which are not soon released, move out and accumulate in large numbers in dilations of the axon.[25] These swellings contain lysosomes which presumably act to destroy the granules.

Theoretically, neurosecretory granules which are not released must either be destroyed in the terminals or be moved out to be broken down elsewhere. Since there is very little evidence of granule degradation in the terminals of many invertebrate neurosecretory systems, one would expect to find preterminal dilations similar to those found in the vertebrate case. Some recent work on neurosecretory axons in aphids has shown that such dilations, packed with granules, do occur.[26]

Maintenance of a pool of releasable hormone

Hormone-releasing systems are subject to discontinuous demand; there are, therefore, problems involved in maintaining a pool of hormone ready for release. In considering these problems, we need first of all to know what constitutes a readily releasable pool. In short-term experiments on the neurohypophysis, it has never been possible to cause the release of more than about one-third of the total hormone content of the neural lobe, and in most cases very much smaller amounts are released. Ultrastructural studies of stimulated and unstimulated neural lobes[27] show clearly why this should be the case; only granules actually in the axon terminals themselves are released and, since the terminals contain only about a third of the total number of neurosecretory granules in the neurohypophysis, obviously no more than this amount can be released unless the pool can be replenished by the arrival of newly synthesized granules from the axons (figure 2.8). This is a slow process compared with the very high rates of hormone release which can be achieved experimentally. Why most experiments result in even lower levels of hormone release has to do with the chain of events linking stimulation to release, and is discussed in the next chapter.

Plainly the size of the pool of hormone in axon terminals depends on three factors: the rate of arrival of granules, the rate of their release, and the rate at which they move into other parts of the axon. A full understanding of the dynamics of this system requires a knowledge of the ways in which each of these three can vary.

A good deal is known about the mechanism and control of granule release. As far as the rate of granule arrival is concerned, however, we have seen (p. 16) that the speed with which granules can be passed by the fast transport system down the axons to the terminals seems not to be capable of much variation. Nonetheless the possibility remains that the rate at which granules arrive at the terminals could be varied, not by changes in the speed with which any of them move, but merely by changes in the number of them in transit. In the last few years it has been found that during stimulation of hormone release there is a concomitant stimulation of

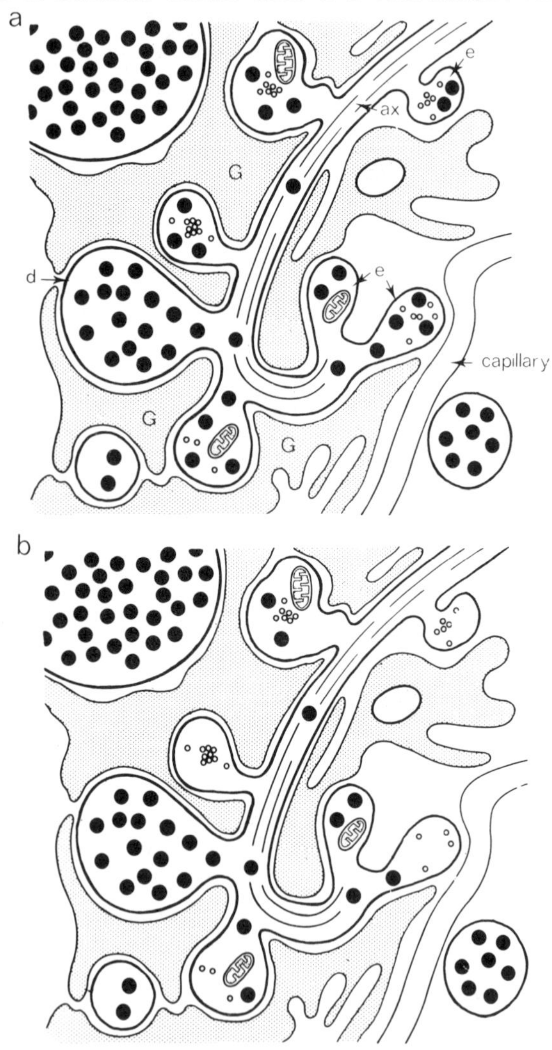

Figure 2.8 The different compartments of the rat neurohypophysis before (*a*), and after (*b*) stimulation of hormone release. Note that, after release, a decrease in the number of neurosecretory granules occurs only in the nerve endings (e), whereas no change can be observed in the dilations or swellings (d) nor in the axons (ax). G, glial cell.

hormone synthesis in the cell body.[28] It seems very likely that this would soon lead to an increase in the rate of arrival of granules at the terminals.

Unfortunately for our understanding of the system, we do not know if there is any regulation of the rate at which neurosecretory granules are removed from the axon terminals into other parts of the axon. The only evidence that there might be any such regulation comes from *in vivo* experiments in which the release of hormone from the neurohypophysis of the dog was stimulated. Although synthesis of new hormone was stimulated during release, there was no correlation between the rate of this synthesis and the hormone depletion of the neural lobe. Furthermore, the accelerated rate of synthesis did not persist after the stimulus was withdrawn.[29,30] As a result, when massive amounts of hormone were

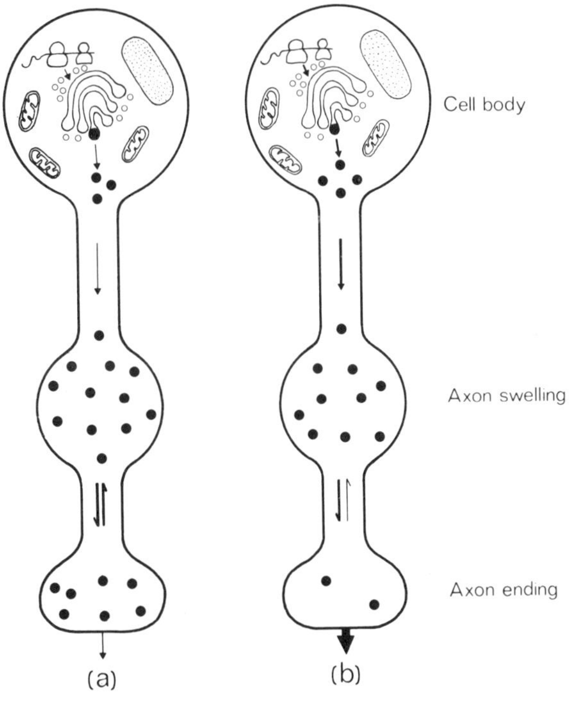

Figure 2.9. A possible way in which an increased rate of hormone release from a neurosecretory cell can be achieved. In (*a*) a low rate of release is seen as involving slow synthesis in the cell body and moderate rates of granule movement down the axon, into the terminal and removal from the terminal. In (*b*) a high rate of hormone release from the terminal is supported by an accelerated rate of granule synthesis in the cell body, moderate rates of movement down the axon and into the terminal, but a sharp reduction in the rate of granule movement out of the terminal.

released in response to an intense stimulus, much less was synthesized than was released. Presumably the only way in which the axon terminals could be rapidly replenished would be by a reduction in the rate at which neurosecretory granules move out into other parts of the system. Whether this occurs or not is not known. There is an interesting problem here for future research. (Figure 2.9 summarizes the flow processes involved in maintaining a supply of hormone-containing granules ready for release.).

Another problem worth mentioning is how the synthesis of new hormone is stimulated when secretion at the nerve terminals is accelerated. Release at the nerve terminals is known to be triggered by the arrival of electrical impulses down the axon. This could well also be the signal for the cell body to accelerate its synthetic activity.

How neurosecretory cells release effective amounts of neurohormones

Perhaps the major difference between nerve cells which release neurotransmitters and those which secrete neurohormones is that the former release their active substance into the very small volume of a synaptic cleft, whereas most hormone secretion is into the vastly more extensive volume of the extracellular space of the whole organism. As an indication of the scale of the difference, it has been calculated that the volume of a synaptic cleft is of the order of 10^{-17} litre whereas for a human, say, the volume of the extracellular fluid accessible to hormones is of the order of 10 litres, i.e. a trillion (10^{18}) times larger. Not surprisingly, neurosecretory cells have become adapted in several different ways to allow them to overcome this difficulty. Apart from a simple increase in the number of cells involved, each neurosecretory neurone has a very large number of axon terminals—in the case of the neurohypophysis, for example, there are about 2000 endings from each cell.

These adaptations achieve a considerable measure of amplification but, in view of the size of the problem, it is to be expected that other adaptations would be needed. One might, for example, predict that neurohormones would be extremely active compounds and that as much hormone as possible would be packed into each neurosecretory granule.

Activity of neurohormones

Neurohormones indeed turn out to be very active compounds. For example, it has been calculated that, in the mouse, oxytocin at concentrations in the range $10^{-12} - 10^{-13}$ M causes contraction of the myometrium of the mammary gland. Similar calculations show that vasopressin acts on tissues sensitive to it at comparable concentrations. In the case of *Rhodnius*, extracts of single neurosecretory cells which synthesize the diuretic hormone of this insect are active when diluted in volumes of fluid 10^8 times as large as themselves.[31] Even if the hormone in the neurosecretory granules is as close-packed as is found in the vertebrate neurohypophysis (see below), it can be calculated that the hormone is active at concentrations at least as low as 10^{-11} M. Because the extracts used in these experiments were of the cell bodies, the neurosecretory granules may well have contained mostly hormone precursor; therefore the hormone itself may be active at much lower concentrations than the calculation above suggests.

These figures can be compared with similar ones for neurotransmitters; acetylcholine is thought to act on post-synaptic membranes at concentrations of around 10^{-9} M.

How neurohormones are packed into neurosecretory granules

From a quantitative ultrastructural study of the neurosecretory granules in the neural lobe of the hypophysis of the rat it has been calculated that the gland contains just over 10^{10} individual granules.[32] Making the assumption that the granules are the only structures to contain active hormone (for which assumption there is now a good deal of evidence, see p. 56) and from a knowledge of the total hormone content of the gland, it can be estimated that each granule contains close to 80 000 hormone molecules.[32] Since the granules have an average diameter of 160 nm, the concentration of hormone in the granules would be somewhat less than 100 mM. At first sight this figure seems low, if it is indeed an advantage for a maximal amount of hormone to be included in each granule. However, it must be remembered that analyses of the contents of neurosecretory granules show that they contain, in addition to active hormone, other proteins (p. 11). We have already pointed out the possible role of the proteins in preventing the neurohormone being lost from the granule by diffusion. If, as we argue above, it is important to pack as much hormone as possible into each granule, such a property would have obvious advantages. It is known, for example, that mixtures of vasopressin and its neurophysin will crystallize out of solution. Whatever the reason for the presence of proteins, the evidence suggests that they are present in the neurohypophysis in equimolar amounts with the hormone. Since the molecular weight of neurophysins is about 10 000 and that of oxytocin and vasopressin each about 1000 one can calculate the approximate minimum space occupied by a molecule of protein-hormone complex[32] to be about 25 nm^3. In a granule of diameter 160 nm only about 80 000 such molecules could be fitted in. In other words, neurosecretory granules from the neural lobe do indeed contain the maximum amount of hormone possible. It is therefore interesting that ultra-structural studies of neurosecretory granules from such diverse animals as the mantis shrimp, blowfly, cat, guinea-pig, hedgehog, ox and rabbit often find them to contain crystalline material.[33] Although it is suspected that these crystals are an artifact of fixation, it seems likely that granule contents are stored in almost solid form in a variety of animal species. An excellent example of this is found in the chromaffin granule.

Chromaffin granules

Although the chromaffin cells of the adrenal medulla represent post-ganglionic sympathetic neurones, they do not have typical neuronal morphology. They have no axons and instead release the hormones they synthesize at the surface of the cell body. The storage granules (chromaffin granules) are so very easily isolated and purified that they have been the subject of a great deal of biochemical and pharmacological work. The results show in some detail just how molecules can be packed and stored into granules. It must be borne in mind, however, that these granules may not be typical of neurosecretory granules, since they store molecules with a molecular weight of only 200 whereas other neurosecretory granules contain peptide hormones whose molecular weights are usually in excess of 1000.

Careful analyses of isolated chromaffin granules by a number of workers have revealed the major components involved and these are listed in Table 2.1.[34, 35, 36] The granules contain high concentrations of catecholamines (*c.* 0.55 M), ATP (*c.* 0.12 M), the divalent cations calcium (20 mM) and magnesium (50 mM), as well as soluble proteins of high molecular weight such as chromogranin A and dopamine-β-hydroxylase. How can such high concentrations be achieved and how can they be kept within the granules? The outer limiting membrane of the granules contains phospholipids, cholesterol and gangliosides[42] at concentrations rather higher than found in other internal membranes. Furthermore, the protein/lipid ratio of the membrane is found to be 0.45 in contrast to a ratio of more than 1 typical of most membranes of cell organelles.

Table 2.1 Composition of the bovine adrenal chromaffin granule.[34]

Constituent	*Amount* *% total dry weight*
Soluble content	
Catecholamines	20.5
Adenine nucleotides	15.0
Protein	27
Calcium	0.1
Magnesium	0.02
Membrane	
Phospholipid	17
Cholesterol	5
Protein	8
Calcium	0.06
Magnesium	0.02

These findings of a high cholesterol but low protein content are characteristic of membranes of low permeability. Such evidence as there is suggests that the granule membrane is indeed rather impermeable to catecholamines and ATP, although these molecules can be moved through the membrane by an energy-requiring process. So, once the granules have been filled with catecholamines and ATP, it is possible that the limited permeability of the membrane could by itself account for the retention of small molecules. However, this is unlikely since it has been shown (figure 2.10) that the granules behave like perfect osmometers, i.e. they respond to

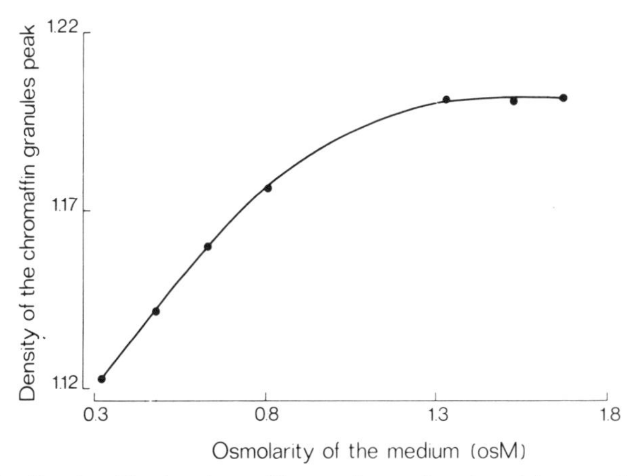

Figure 2.10. Density of bovine chromaffin granules as a function of the osmotic pressure of the suspension medium (modified from Morris & Schovanka, ref. 37).

changing osmotic concentrations as one would expect for a volume surrounded by a semi-permeable membrane and containing a constant number of osmotically active particles in solution (figure 2.10) To act in this way as an osmometer means that the osmotic concentration inside the granule must match that of the external solution. Therefore at physiological osmolalities (~ 300 mosM) it follows that only a small fraction of the total content of catecholamines and ATP can be dissolved in aqueous solution, the rest must exist in some osmotically inactive form.

Using bovine adrenal medulla chromaffin granules, it has been possible to observe changes in granule density and refractive index in response to changing osmolality of the bathing solution.[37] From the sizes of these changes it can be calculated how much of the granules is occupied by a core of osmotically inactive material and how much by aqueous solution. As figure 2.11 shows, to explain the results, about 37% of the granules must be occupied by condensed core material containing the bulk of the catecholamine, ATP and protein content.[38]

Experiments with solutions of catecholamines, ATP and Ca^{2+} and/or Mg^{2+} have shown that when these components are mixed at concentrations similar to those found in the granules, they can form high-molecular-weight aggregates which separate from solution as a viscous fluid containing very little water. It is possible then that just such a process occurs in the granule *in vivo*. At the moment it seems likely that

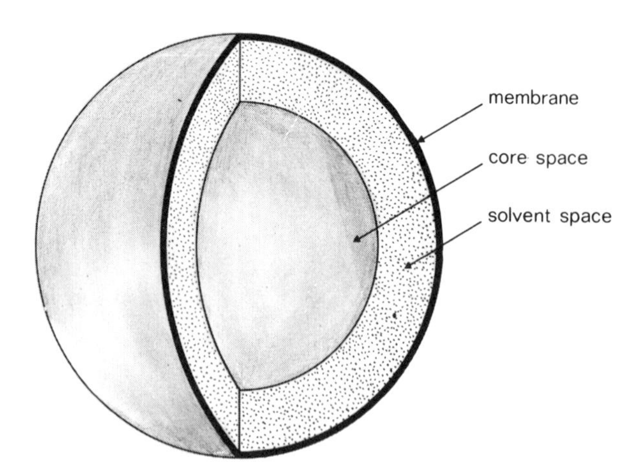

Figure 2.11. Model for the structure of an isolated bovine adrenal medulla chromaffin granule. The contents of the core and the solvent space are in equilibrium (from data of Morris *et al.*, ref. 38).

chromogranin A, the protein found in the granule, does not itself bind more than small amounts of catecholamines. It is speculated that the role of the chromogranin might instead be to provide a physical environment to stabilize aggregates of ATP and catecholamine. High concentrations of isolated chromogranin A will spontaneously form a gel *in vitro*, and such a gel *in vivo* might well influence the dissociation of ATP-catecholamine complex. In addition, since chromogranin carries many negative charges, it may slow down diffusion of such small charged molecules as catecholamines.[34]

We may summarize the structure of the granules as follows. High concentrations of catecholamines and ATP in the granule interact in the presence of calcium and magnesium ions to form high-molecular-weight aggregates in an osmotically inert non-aqueous phase which is stabilized by the chromogranin A. The net dissociation of these aggregates is prevented by the maintenance of an equilibrium with these constituents in solution within the granule. The materials in solution are maintained by pumps in the granule membrane which replenish the catecholamine and ATP. These pumps use ATP as an energy source and have properties similar to proton translocating ATP-ases found in mitochondria.[39]

We have seen in this chapter how neurosecretory cells release enough of their active products to have an effect when they are diluted in the extracellular fluid. Essentially they do this by synthesizing very active material, by packing the maximum amount of this material into each granule, and by releasing them simultaneously from a very large number of axon endings.

REFERENCES

1. Gainer, H., Loh, Y. P. & Sarne, Y. (1977) "Biosynthesis of neuronal peptides," Ch. 8 in *Peptides in Neurobiology* (ed. H. Gainer), New York, Plenum Press.
2. Sachs, H. & Takabatake, Y. (1964) "Evidence for a precursor in vasopressin biosynthesis," *Endocrinology* **75**, 943–948.
3. Pickering, B. T., Jones, C. W., Burford, G. D., McPherson, M., Swann, R. W., Heap, P. F. & Morris, J. F. (1975) "The role of neurophysin proteins: suggestions from the study of their transport and turnover," *Ann. N.Y. Acad. Sci.* **248**, 15–35.
4. Cross, B. A., Dyball, R. E. J., Dyer, R. G., Jones, C. W., Lincoln, D. W., Morris, J. F. & Pickering, B. T. (1975) "Endocrine neurons," *Rec. Prog. Horm. Res.* **31**, 243–286.
5. Gainer, H., Sarne, Y. & Brownstein, M. J. (1977) "Neurophysin biosynthesis: conversion of putative precursor during axonal transport," *Science, N.Y.* **195**, 1354–1356.
6. Sachs, H., Fawcett, P., Takabatake, Y. & Portanova, R. (1969) "Biosynthesis and release of vasopressin and neurophysin," *Rec. Prog. Horm. Res.* **25**, 447–491.
7. Valtin, H., Sokol, H. W. & Sunde, D. (1975) "Genetic approaches to the study of the regulation and actions of vasopressin," *Rec. Prog. Horm. Res.* **31**, 447–486.
8. Sunde, D. A. & Sokol, H. W. (1974) "Quantification of rat neurophysins by polyacrylamide

gel electrophoresis (PAGE): application to the rat with hereditary hypothalamic diabetes insipidus," *Ann. N.Y. Acad. Sci.* **248,** 345–364.

9. Cannata, M. A. & Morris, J. F. (1973) "Change in the appearance of hypothalamo-neurohypophysial neurosecretory granules associated with their maturation," *J. Endocr.* **57,** 531–538.

10. Girardie, J. & Girardie, A. (1972) "Evolution de la radioactivité des cellules neurosécrétrices de la pars intercerebralis chez *Locusta migratoria migratoroïdes* (Insecte Orthoptère) après injection de cystéine S.[35]. Étude autoradiographique aux microscopes optique et électronique," *Z. Zellforsch. Mikrosk. Anat.* **128,** 212–226.

11. Maddrell, S. H. P. (1974) "Neurosecretion," Ch. 6 in *Insect Neurobiology* (ed. J. E. Treherne), Amsterdam, Oxford, North Holland.

12. Aston, R. H. & White, A. F. (1974), "Isolation and purification of the diuretic hormone from *Rhodnius prolixus*," *J. Insect Physiol.* **20,** 1673–1682.

13. Eckert, M. (1973) "Immunologische Untersuchungen des neuroendokrinen Systems von Insekten. III. Immunohistochemische Markierung des neuroendokrinen Systems von *Periplaneta americana* mit durch Fraktionerung von Retrocerebralcomplex-extrakten gewonnen Anti-Seren," *Zool. Jb. Physiol.* **77,** 50–59.

14. Gabe, M. (1972) "Histochemical data on the secretion of protocephalic neurosecretions in pteryote insects during axonal transport," *Acta Histochem. (Jena)* **43,** 168–183.

15. Mordue, W. & Goldsworthy, G. J. (1969) "The physiological effects of corpus cardiacum extracts in locusts," *Gen. comp. Endocrinol.* **12,** 360–369.

16. Maddrell, S. H. P. (1962) "A diuretic hormone in *Rhodnius prolixus* Stål, *Nature, Lond.* **194,** 605–606.

17. Maddrell, S. H. P. (1966) "The site of release of the diuretic hormone in *Rhodnius*—a new neurohaemal system in insects," *J. exp. Biol.* **45,** 499–508.

18. Rodriguez, E. M. & Dellmann, H. D. (1970) "Ultrastructure and hormonal content of the proximal stump of the transected hypothalamoneurohypophysial tract of the frog (*Rana pipiens*)," *Z. Zellforsch. Mikrosk. Anat.* **104,** 445–479.

19. Harker, J. E. (1960). "Endocrine and nervous factors in insect circadian rhythms," *Cold Spring Harb. Symp. quant. Biol.* **25,** 279–287.

20. Heslop, J. (1975) "Axonal flow and fast transport in nerves," *Adv. Comp. Physiol. Biochem.* **6,** 75–163.

21. Grafstein, B. (1977) "Axonal transport: the intracellular traffic of the neuron," in *Handbook of Physiology*, Section 1. "The nervous system," Bethesda, Amer. Physiol. Soc.,691–717.

22. Jasinski, A., Gorbman, A. & Hara, T. J. (1966) "Rate of movement and redistribution of stainable neurosecretory granules in hypothalamic neurons," *Science, N.Y.* **154,** 776–778.

23. Highnam, K. C. & Mordue, A. J. (1970) "Estimates of neurosecretory activity by an autoradiographic method in adult female *Schistocerca gregaria* (Forsk.)," *Gen. comp. Endocr.* **15,** 31–38.

24. Steel, C. G. H. & Harmsen, R. (1971) "Dynamics of the neurosecretory system in the brain of an insect, *Rhodnius prolixus*, during growth and molting," *Gen. comp. Endocr.* **17,** 125–141.

25. Heap, P. F., Jones, C. W., Morris, J. F. & Pickering, B. T. (1975) "Movement of neurosecretory product through the anatomical compartments of the neural lobe of the pituitary gland," *Cell Tiss. Res.* **156,** 483–497.

26. Steel, C. G. H. (1977) "The neurosecretory system in the aphid *Megoura viciae*, with reference to unusual features associated with long distance transport of neurosecretion," *Gen. Comp. Endocrinol.* **31,** 307–322.

27. Nordmann, J. J. & Morris, J. F. (1976) "Membrane retrieval in neurosecretory axon endings," *Nature* **261,** 723–725.

28. Takabatake, Y. & Sachs, H. (1964) "Vasopressin biosynthesis. III. *In vitro* studies," *Endocrinology* **75,** 934–942.

29. Picard, D., Michel-Brechet, M., Athouël, A. M. & Rua, S. (1972) "Granules neurosécrétoires, lysosomes et complexe GRL dans le noyau supra-optique du rat. Bipolarité des complexes golgiens," *Exptl. Brain Res.* **14,** 331–354.
30. Morris, J. F. & Dyball, R. E. J. (1974) "A quantitative study of the ultrastructural changes in the hypothalamo-neurohypophysial system during and after experimentally induced hypersecretion," *Cell Tiss. Res.* **149,** 525–535.
31. Maddrell, S. H. P. (1963) "Excretion in the bood-sucking bug, *Rhodnius prolixus* Stål. I. The control of diuresis," *J. exp. Biol.* **40,** 247–256.
32. Morris, J. F. (1976) "Hormone storage in individual neurosecretory granules of the pituitary gland: a quantitative ultrastructural approach to hormone storage in the neural lobe," *J Endocr.* **68,** 209–224,
33. Normann, T. C. (1976) "Neurosecretion by exocytosis," *Int. Rev. Cytol.* **46,** 1–77.
34. Winkler, H. & Smith, A. D. (1975) "The chromaffin granule and the storage of catecholamines," in *Adrenal Gland*, Handbook of Physiology (eds. Blaschko, H., Sayers, G. & Smith, A. D.), Vol. 6, pp. 321–339, Bethesda: Amer. Physiol. Soc.
35. Smith, A. D. & Winkler, H. (1972) "Fundamental mechanisms in the release of catecholamines," in *Catecholamines*, Handbook of Experimental Pharmacology (eds. Blashko, H. & Muschall, E.), Vol. 33, pp. 538–617, Berlin, Springer Verlag.
36. Winkler, H. (1976) "The composition of adrenal chromaffin granules: an assessment of controversial results," *Neuroscience* **1,** 65–80.
37. Morris, S. J. & Schovanka, I. (1977) "Some physical properties of adrenal medulla chromaffin granules isolated by a new iso-osmotic density gradient procedure," *Biochim. Biophys. Acta* **464,** 53–64.
38. Morris, S. J., Schultens, H. A. & Schober, R. (1977) "An osmometer model for changes in the buoyant density of chromaffin granules," *Biophys. J.* **20,** 33–48.
39. Bashford, C. L., Casey, R. P., Radda, G. K. & Ritchie, G. A. (1976) "Energy-coupling in adrenal chromaffin granules," *Neuroscience* **1,** 399–412.
40. Swaab, D. F., Nijveldt. F. & Pool, C. W. (1975) "Distribution of oxytocin and vasopressin in the rat supraoptic and paraventricular nucleus," *J. Endocr.* **67,** 461–462.
41. Vandesande, F. & Dierickx, K. (1975) "Identification of the vasopressin producing and of the oxytocin producing neurons in the hypothalamic magnocellular neurosecretory system of the rat", *Cell. Tiss. Res.* **164,** 153–162.
42. Dreyfus, H., Aunis, D., Edel, S. and Mandel, P. (1977) "Gangliosides and phospholipids of the membranes from bovine adrenal medulla chromaffin granules", *Biochim. Biophys. Acta,* **489,** 89–97.

CHAPTER THREE

ELECTRICAL PROPERTIES OF NEUROSECRETORY CELLS

THE LAST CHAPTER DESCRIBED HOW NEUROSECRETORY CELLS SYNTHESIZE neurohormones, package them into membrane-bound granules, and transport them down the axons to their endings where they accumulate ready for liberation. In this chapter we shall show that neurosecretory cells have all the bioelectrical properties that are found in other non-endocrine neurones. Also we shall describe the patterns of impulse production in neurosecretory cells, how they are produced, and what factors affect this behaviour. Finally we shall see how changes in the bioelectrical activity of neurosecretory cells are correlated with changes in the rate of hormone release, and look at the evidence that electrical activity might affect the rate of synthesis of hormone. The upshot of these findings is that it is believed that hormone release is triggered by impulse traffic in the neurosecretory axons and that changing rates of hormone release are produced by changes in the pattern and intensity of the impulse flow.

Bioelectrical activity of neurosecretory cells

Although neurosecretory cell bodies are not randomly situated in the central nervous system, they are intermingled to some extent with the cell bodies of non-endocrine neurones; so, before one can investigate the electrical activity of neurosecretory cells, one has to be able to identify the cells which are neurosecretory. This identification is usually achieved either electrically or histologically.

The electrical method relies on the fact that while the cell bodies may be surrounded by other (non-endocrine) cells, their axon endings are usually segregated. As we have seen earlier, the axon terminals of neurosecretory cells commonly lie at some distance from their respective cell bodies. In some animals, the terminals are closely packed together; as a result, the aggregated endings can easily be seen, and access to them gained by relatively simple dissection. For example, the axons of neurosecretory cell bodies lying in the supraoptic and paraventricular nuclei of the brain of mammals collect together and run out from the base of the brain as the *infundibular stalk*; the axons then all branch and end as a swollen organ called the neurohypophysis (see p. 98). Now nerve axons can be made to conduct impulses either in the normal direction from cell body to terminals (*orthodromically*) or in the opposite direction (*antidromically*). Neurosecretory axons are no exception to this rule. Therefore, if one stimulates the neurohypophysis or the infundibular stalk, antidromic impulses will run back up the neurosecretory axons to their cell bodies (figure 3.1). As a result, by making recordings in the brain, those cell bodies which respond can be found. There are, however, precautions one has to take before it can be concluded that such responding cell bodies are neurosecretory. There is the possibility that a recurrent branch of the stimulated neurosecretory axon might synapse with another cell and so stimulate it orthodromically. To prove in a particular case that this is not so, the appearance of an impulse initiating in the cell body (either spontaneously or stimulated there) is used as a trigger for simultaneous artificial stimulation of the axon endings. If the antidromic spike subsequently fails to appear in the cell body, then both spikes must have been in the same cell (figure 3.1). The reasoning behind this is that the only way an antidromic spike will fail to reach the cell body is if the axon is in a refractory state following the passage of an impulse, in this case the orthodromic spike.

These sorts of experiments have been done, for example, in the goldfish,[1] cat,[2] rabbit,[3] and rat[4] and the cells thus revealed as neurosecretory turn out to be most densely situated in those areas which histological examination shows to contain large numbers of neurosecretory cells.

Where intracellular techniques can be applied, not only can absolute identification of a given cell body as being neurosecretory usually be made, but there is the further advantage that a wide range of the electrical properties of the cell can be explored in detail. The neurosecretory cells of the goldfish brain, for example, are large enough to allow easy penetration by intracellular microelectrodes.[1] It was work on this animal which provided a good deal of the basis for what we now know of the electrical properties of neurosecretory cells.

The alternative method of identifying cells as neurosecretory is histological, using the staining properties or ultrastructural appearance of the cells. For example, neurosecretory cells in the molluscan central nervous system are readily identified in this way. Many of them are very large (100–500 μm in diameter) so their electrical behaviour can very conveniently be studied even by methods which involve the placing of more than one microelectrode in the cell. Since they are often characteristically

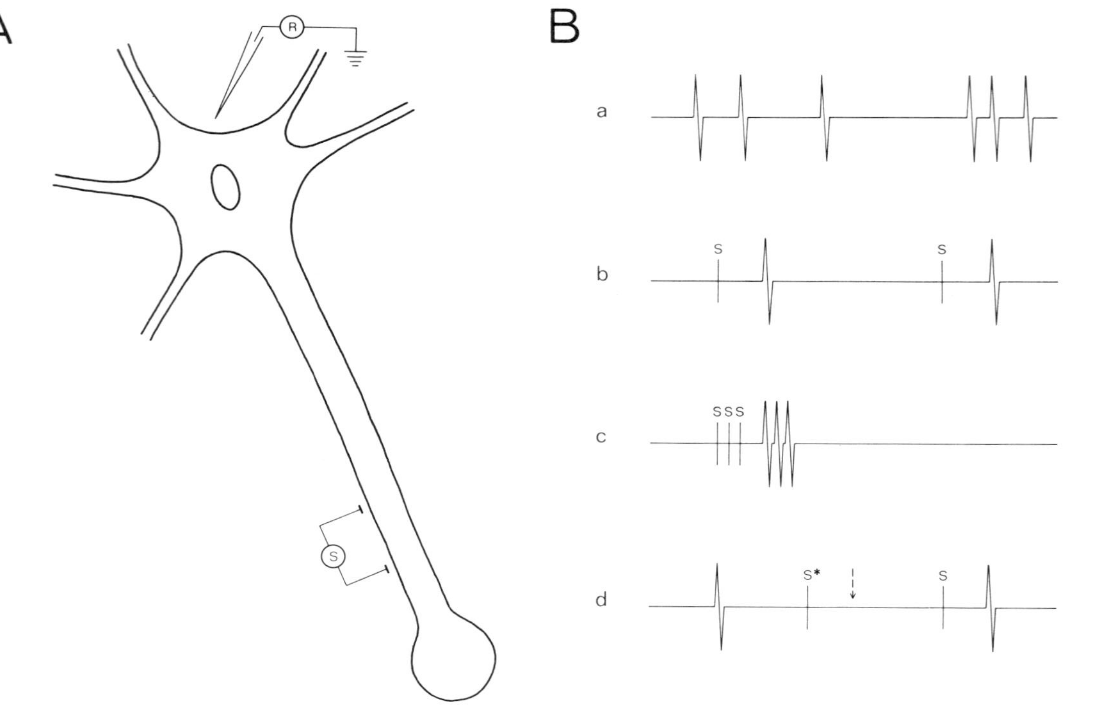

Figure 3.1 Electrical identification of hypothalamic neurosecretory cell bodies. A. Diagrammatic representation of the positions of recording and stimulating electrodes. R = extracellular recording electrode. S = stimulating electrodes on the pituitary stalk. B. electrical recordings from an extracellular electrode sited close to a cell body. (*a*) A recording from a spontaneously active cell (*b*) Recordings of antidromic spikes evoked by stimuli at the times indicated (*s*). Note the constant interval (latency) between the stimulus and the arrival of the spike. (*c*) As (*b*) but showing the repetitive response to a train of stimuli (delivered at *sss*). Again the latency is constant. (*d*) At *s**, an orthodromic spike triggers an antidromic spike which fails to reach the cell body (at R) because the cell membrane is in a refractory state due to the orthodromic spike. Later, at *s*, an antidromic spike is evoked which is recorded at the cell body.

pigmented, there is the additional advantage that the same cell can be studied in different animals. As a result, work on animals in different physiological states gives some idea as to what are biologically significant changes in electrical behaviour.

What do recordings from neurosecretory cells show? So far, the investigations which have been made all agree that neurosecretory cells have bioelectrical properties which in most respects are very similar to those of non-endocrine nerve cells. Their resting membrane potentials are generally in the range 40–60 mV and they produce action potentials both spontaneously, and when stimulated by depolarizing current pulses. In the snail,[5] sea hare[6] and the goldfish,[1] for example, the action potentials have an overall amplitude of between 80 and 120 mV. A typical neurosecretory action potential is shown in figure 3.2. In recordings from the cell bodies of molluscan neurosecretory cells, the initial, inward current is carried by both Na^+ and Ca^{2+} ions. In the snail *Helix*, for example, the size of the action potential declines by 30% in a calcium-free solution and is abolished in a saline lacking both sodium and calcium (figure 3.3). In these cell bodies, then, an inward current of either ion is sufficient for the production of action potentials, albeit of reduced size. This is in contrast to most axonal action potentials, of course, where sodium ions alone carry the inward current, so that conduction rapidly fails in sodium-free solutions.

One apparent difference between the electrical behaviour of neurosecretory cells and non-endocrine nerve cells is that neurosecretory cells, particularly those of invertebrates, tend to have action potentials of greater duration. In non-neurosecretory cells the action potential generally lasts less than 1 ms, whereas most neurosecretory cells have action potentials in

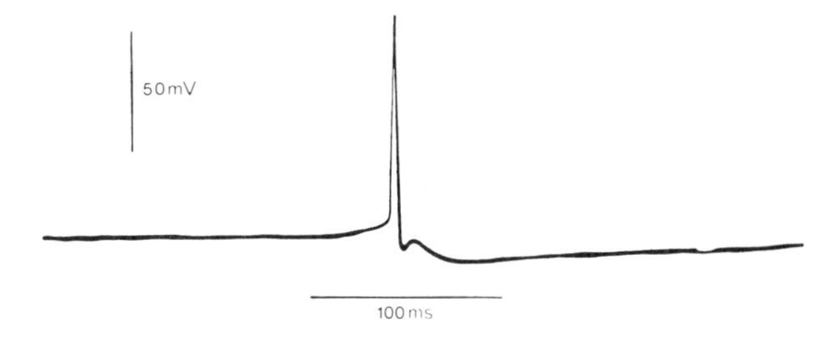

Figure 3.2 Intracellular recording of an action potential of a hypothalamic neuroendocrine cell of the goldfish (redrawn from Kandel, 1964, ref. 1).

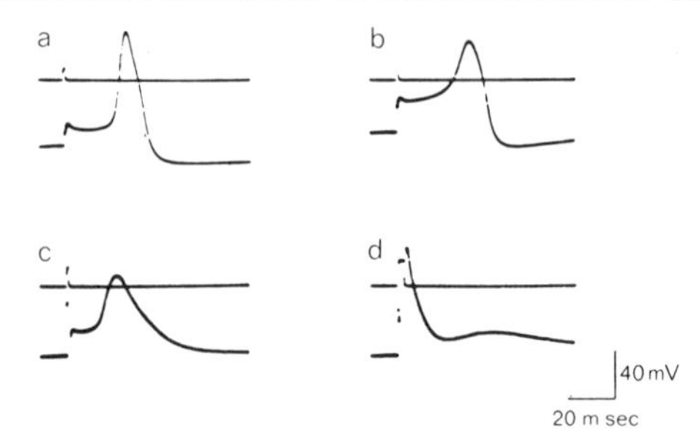

Figure 3.3 Intracellular recording from a snail neuroendocrine cell body: the effects of sodium-free and calcium-free solutions. The action potentials were induced by brief electrical pulses. (a) normal saline; (*b*) after 30 min in sodium-free solution; (*c*) after 10 min in calcium-free saline; (*d*) (Na,Ca)-free saline. The upper trace marks the zero level and the stimulating current pulse (after Standen, 1975, ref. 5).

the range 5–10 ms with some as long as 20 ms. In a few cases, as in the sinus gland of crabs, it has been possible to make direct intracellular recordings from the very large neurosecretory axon endings[7] (figure 3.4). In these endings the action potential following the arrival of an impulse down the axon is also prolonged. From extracellular recordings made at the surface of the sinus gland it has, however, been possible to deduce that the axonal action potentials have an ordinary, relatively short, time course, but that these are followed by much slower action potentials in the axon terminals. The inward current of these latter action potentials is carried by both sodium and calcium ions (figure 3.4) and it has been speculated that the unusual length of the action potential is a local adaptation to permit prolonged entry of extracellular calcium ions which, it is now known, play a central role in promoting hormone release. It is relevant to point out that in the synapses of the squid stellate ganglion, the amount of transmitter released increases with the duration of the depolarization of the presynaptic membrane.

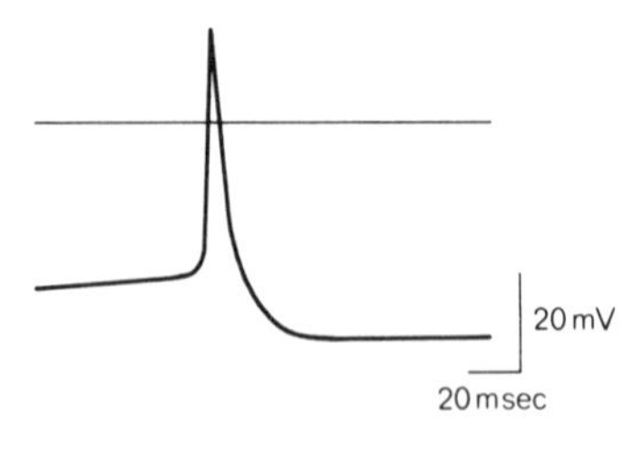

Figure 3.4 Intracellular recording from a neurosecretory axon ending in the sinus gland of the crab *Cardisoma guanhumi* (courtesy of I. M. Cooke).

Bioelectrical activity and hormone release

As we have just discussed, neurosecretory cells have electrical properties similar in most respects to those of conventional nerve cells. The overwhelming weight of evidence is now that they release their hormones by a process triggered by a flow of impulses along their axons. Early support for this was that direct electrical stimulation of neurosecretory structures would often cause hormone release in amounts related to the intensity of the stimulation. In addition, as we shall discuss later, it is known that treatments that mimic the electrical events thought to occur in nerve endings, can lead to massive release of hormone.

The best evidence that impulse traffic and hormone release are related comes from recent work on vasopressin-secreting cells in the rat (figure 3.5). The evidence is of two parts. First, during experiments in which the plasma osmotic pressure was increased by injection of hyperosmotic saline, there was an almost exactly proportional increase in circulating level of vasopressin.[8] Incidentally, it was estimated that even a 1% increase in osmotic concentration was sufficient to produce a significant change in the concentration of vasopressin in circulation.[8] Secondly, such injections caused an increase in the firing of neurosecretory cells in the paraventricular and supraoptic nuclei.[4, 9, 10] Similar increases in electrical activity follow chronic dehydration and, significantly, the increase in firing is almost linearly proportional to the increased osmotic concentration of the plasma.[12] These results provide the basis for the belief that vasopressin secretion is brought about by nerve impulses in the cells, and that increases in the rate of liberation of hormone are achieved by an intensification of

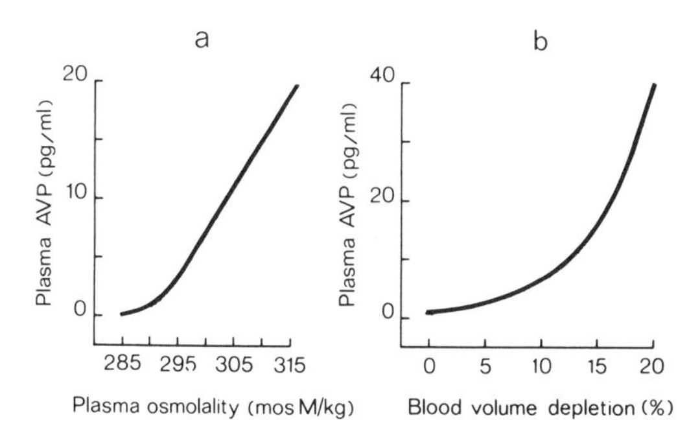

Figure 3.5 (*a*) Changes in plasma vasopressin (AVP) in response to changes in plasma osmolality. (*b*) Changes in plasma vasopressin (AVP) in response to iso-osmotic changes in blood volume (after Dunn *et al.*, 1973, ref. 8).

the cells' electrical activity. The so-called Brattleboro strain of rats provides an interesting piece of confirmatory evidence. In these animals the neurosecretory cells of the brain are unable to synthesize vasopressin and their firing rate is permanently elevated.[13]

A criticism of this type of work has been that the electrically active neurones of the supraoptic and paraventricular nuclei were identified as neurosecretory on the basis of their response to antidromic stimulation from the neurohypophysis. They may, therefore, have included oxytocin as well as vasopressin cells. However, vasopressin cells can be distinguished from oxytocin cells by the fact that, unlike oxytocin cells, they do not respond to treatments which provoke milk ejection by causing release of oxytocin (see also p. 141). Experiments have now been done on the response of such identified vasopressin cells to haemorrhage, a known stimulus for vasopressin release. The results (figure 3.6) showed that a reduction in blood volume sufficient to cause vasopressin release, elicits an increased electrical activity in the cells.[14] The electrical activity is most intense in cells of animals subject to the most intense stimuli for vasopressin release. When blood is replaced into animals with initially reduced blood volumes, the

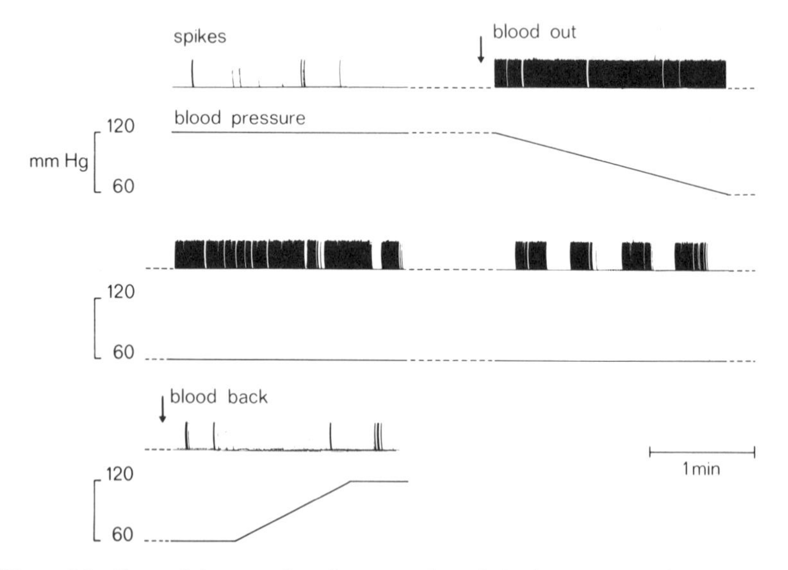

Figure. 3.6 Extracellular recordings from a rat hypothalamic neuroendocrine cell during haemorrhage. Note that blood withdrawal leads to a decrease in blood pressure and an increase in firing rate of the cell. The cell then fired continuously for a while before the firing broke up into "bursts", the phasic pattern. When blood was reinjected, the cell's firing returned to its original pattern (after Poulain *et al.*, ref. 14).

electrical activity of the vasopressin cells returns to the slow discharge rate characteristic of cells in osmotic balance with a normal blood volume.

The upshot of this work is that electrical activity in the cells and hormone release are closely correlated. Strictly speaking this does not prove that the one causes the other, but it is a reasonable working hypothesis that this is the case.

Firing patterns of neurosecretory cells and their significance

During the work described above a striking feature emerged. As the electrical activity of the vasopressin cells increased, the pattern of firing changed from a slow irregular discharge, not to a regular rapid discharge, but to one in which impulses were generated in bursts at a much higher frequency with short quiescent periods in between (figure 3.6). What might be the significance of this type of firing pattern?

One attractive explanation is that during a volley of impulses there might be facilitation of hormone release. That facilitation does occur in release mechanisms of the nervous system has been shown, for example, from experiments on crayfish motor neurones.[15] In this case a change in firing from a regular discharge to "bursting" behaviour with *no* increase in the overall spike frequency led to an increased tension in the muscle, clear evidence of an increase in the release of neurotransmitter. No quite such direct evidence is available for the release of neurohormones, but there is some indirect evidence of facilitation. In these experiments, isolated rat neurohypophyses were stimulated electrically at various frequencies and the quantities of oxytocin released were measured. A given number of pulses was found to release progressively more hormone as they were given at higher frequencies (figure 3.7); in other words, hormone release depended directly on the frequency of stimulation.[16] This being the case, it is easy to believe that volleys of impulses would release more hormone than the same numbers of impulses arriving more uniformly spaced.

Although such bursting behaviour is not unique to neuroendocrine cells, it is an important feature of many of them, so that it is worth considering how such firing activity might be produced.

One obvious possibility is that there might be oscillatory changes in the synaptic stimulation that the neuroendocrine cells receive. They could be made silent by inhibitory input involving the development of a hyperpolarizing post-synaptic potential, or be made to increase their firing rate by excitatory synapses causing depolarization of the cell membrane. In

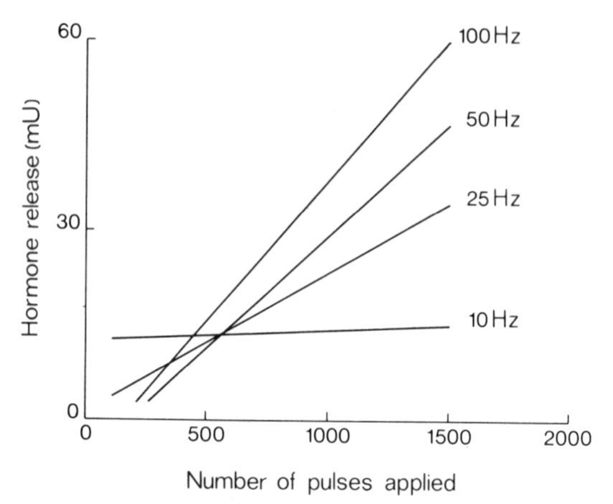

Figure 3.7 The relation between frequency of stimulation and hormone release from rat neurohypophyses incubated *in vitro* and stimulated electrically (after Nordmann & Dreifuss, 1972, ref. 16).

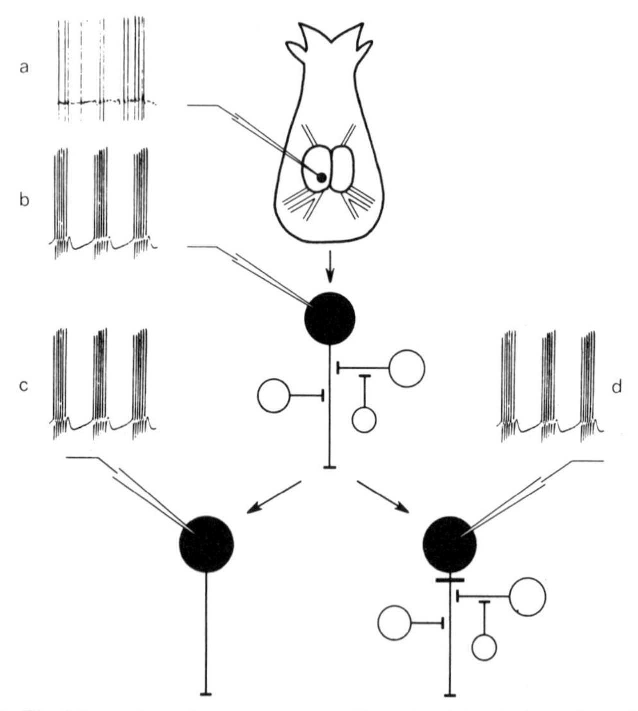

Figure 3.8 The firing pattern of neurosecretory cells in the abdominal ganglion of *Aplysia*. (*a*) shows the behaviour of a cell from a ganglion still in place in the animal. (*b*) the rhythmic bursting pattern of a cell in an isolated ganglion. (*c*) and (*d*) bursting firing of a neurosecretory cell body isolated either surgically (*c*) or by ligature (*d*).

the goldfish, for instance, stimulation of the olfactory tract gives rise to graded excitatory post-synaptic potentials in the hypothalamic neurosecretory cells. Furthermore, antidromic spikes led not only to impulses arriving in the cell bodies but also to the production of inhibitory post-synaptic potentials possibly through a collateral inhibitory system.[1]

However, although such synaptic action of other cells is undoubtedly of great importance in controlling the electrical activity of neurosecretory cells, recent experiments have shown that at least some hormone-releasing cells can show intrinsic patterns of activity (figure 3.8). For instance, in *Aplysia* it has been possible by ligature to isolate single neurosecretory cells from the rest of the cell population, and they still continue to produce action potentials in a bursting pattern.[17] In the land snail *Otala*, similar results have been obtained from a cell body together with a short length of its axon which were completely removed from the ganglion and studied in isolation.[18] In fact, in these two molluscs, recordings made from neurosecretory cells in ganglia still *in situ* in the animal do not show quite such clear bursting behaviour.[19] The cells are clearly under synaptic control, and their behaviour becomes regulated by the activity of the rest of the central nervous system.

The mechanism underlying the rhythmic activity of neurosecretory cells

Since the intrinsic bursting activity described above represents behaviour which is not at once explicable on the known basis of the resting potential and action potential of, say, the squid giant axon, and because of its importance, it is worth giving some idea of what special features are thought to be involved in nerve cells which exhibit such rhythmic behaviour.

An explanation which can satisfactorily account for the observations is that the membrane potential is driven up and down in a rhythmic way by changes in the potassium conductance of the surface membrane. This, in turn, is consequent on changes in the calcium concentration of the cytoplasm. When the membrane is in the depolarized state, action potentials are generated in the normal way and, when hyperpolarized, the cell becomes quiescent as one would expect. Basically then nerve cells which are intrinsically rhythmically active differ from other nerve cells in that their membrane potential oscillates.

For bursting neurosecretory cells, the evidence for calcium-dependent changes in potassium conductance is good; intracellular injections of calcium-containing solutions lead to rapid increases in potassium

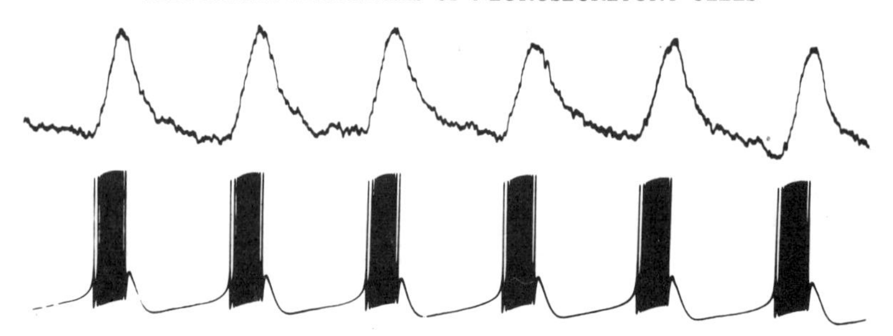

Figure 3.9 Intracellular recording and free intracellular calcium ion concentration measurement in a rhythmically active neurosecretory cell from *Aplysia*. The upper trace illustrates the increases in cytoplasmic calcium concentration associated with the spontaneous bursts of action potentials (lower trace). The changes in free internal Ca^{2+} concentration were measured from changes in absorbance of the dye, Arsenazo III (after Gorman & Thomas, 1978, ref. 22).

conductance.[20,21] In addition, recent experiments have shown that intracellular calcium levels in rhythmically active neurosecretory cells vary in phase with the bursts of action potentials[22] (figure 3.9). What is not yet clear is why the intracellular concentration of calcium ions should vary in an oscillatory fashion. One suggestion is that the permeability of the cell membrane to calcium ions is potential-dependent. The chain of events would then be as follows (figure 3.10). During depolarization, the calcium

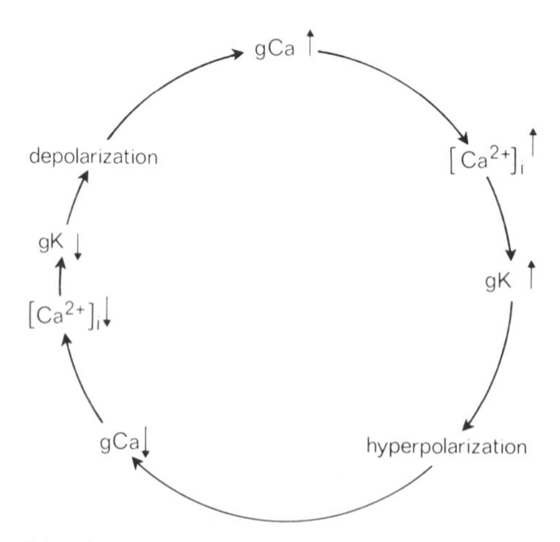

Figure 3.10 Possible scheme to explain the oscillatory changes in membrane potential of rhythmically active nerve cells.

permeability of the cell membrane would increase, and calcium would enter the cell from the external medium (resting intracellular levels of calcium are very low, in the range 10^{-8} – 10^{-7} M). The increase in intracellular calcium level would then lead to an increase in the potassium conductance of the membrane, causing the membrane to hyperpolarize. In turn this would lead to a decrease in calcium permeability, which would allow the extra calcium which had entered to be removed, either by being bound intracellularly and/or by being pumped out from the cell. The resulting fall in calcium level would lead to a decrease in potassium conductance of the membrane and so to a depolarization again. In this way, the membrane potential and the conductances of the membrane to calcium and potassium would be linked in a feedback loop. With sufficient delays in the feedback effects, oscillations in each of the contributory elements will ensue.

In some cases it appears that the system does not rely on calcium entry from outside the cell. To account for this, the interesting suggestion has been made[23] that oscillations in intracellular levels of calcium might stem from a feedback loop involving the intracellular levels of cyclic AMP,[24] which are known to affect the release of calcium from intracellular stores.

It must be emphasized that the preceding explanations are not the only ones which exist. Indeed, it is certain that a variety of other effects affect the form and frequency of the oscillations in membrane potential. For example, changes in membrane sodium conductance, electrogenic effects of the Na/K pump, and intracellular pH all alter the rhythmic electrical behaviour of bursting nerve cells.[25] However, while some theories involve one or more of these effects as of central importance, it is possible to believe that they play only a secondary role in modulating a primary oscillation in potential induced by changes in the cellular levels of calcium.[26] Whether this view is a correct one will emerge from the very active research now going on in this interesting area.

Circadian, monthly and annual changes in electrical activity of neurosecretory cells

Before we leave the subject of rhythmically active neurosecretory cells it is of great interest to note that some snail neurosecretory neurones have been found, not only to fire rhythmically, but in which the pattern changes during the day in a regular fashion, i.e. a diurnal or circadian rhythm occurs in the firing rate. In the land snail *Otala*, for instance, one particular neurone was found to fire in bursts for most of the day but, around midday, the cell firing pattern changed to a continuous train of spikes (figure 3.11).

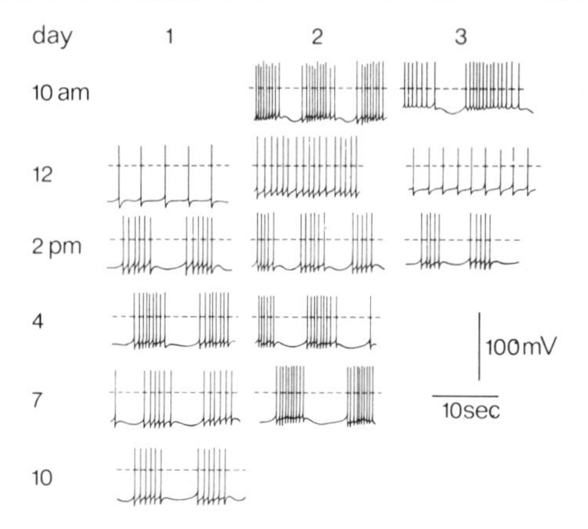

Figure 3.11 Sample records from 3 days of intracellular recording from a neuroendocrine cell of the snail *Otala lactea*. Note the circadian rhythm in spike pattern evident over the 3 days. The cell was predominately a bursting pacemaker, but exhibited beating pacemaker activity around 12 noon. By 2 p.m. the cell was again in a burst mode and remained so throughout the evening (data not shown) (adapted from Gainer, 1972, ref. 18).

Furthermore, during the winter months November to March, when the snail goes into an inactive state (diapause), this cell became electrically inactive, while other cells showed no change in their behaviour.[18] It is possible, therefore, that the neurosecretory activity of this cell may be related to the control of the animal's activity.

As an additional complexity, in the sea hare *Aplysia*, the circadian firing pattern of at least one neurosecretory cell is modulated so that the time of day at which the firing activity is greatest, moves in time with the 14-day lunar tidal cycle.

Control of bioelectrical activity of neurosecretory cells

We have seen on p. 41 some evidence showing that neurosecretory cells can be affected by inhibitory and/or excitatory synapses. In the case of inhibition following antidromic stimulation, it is thought that this occurs through a collateral inhibitory system—in other words a branch of the axon is thought either to run back and end at the synaptic area of the same cell or to synapse with one (or more) interneurones which then synapse with the neurosecretory cell. It has been suggested that the inhibitory transmitter substance released by the collateral branch might be some of the hormone which is liberated at the main release sites. However, in experiments using the Brattleboro strain of rats which are congenitally completely unable to synthesize vasopressin, the vasopressin cells still show recurrent inhibition,[27] so either the transmitter substance is not vasopressin or perhaps the proposed recurrent inhibitory axon does not exist. It will be interesting to discover just how the inhibition is brought about.

In addition to synaptic control of neurosecretory cells, there is also evidence for humoral control. The first suggestion of this was the finding that vertebrate peptide hormones such as oxytocin and vasopressin would cause long-lasting changes in the pattern of firing in some molluscan neurosecretory cells[28] (figure 3.12). Since they were active at concentrations as low as 10^{-9} M, and since many other vertebrate hormone and releasing factors were without effect, it seemed that their action was a relatively specific one. The effect produced by these hormones was to cause an initiation or potentiation of the production of impulses in

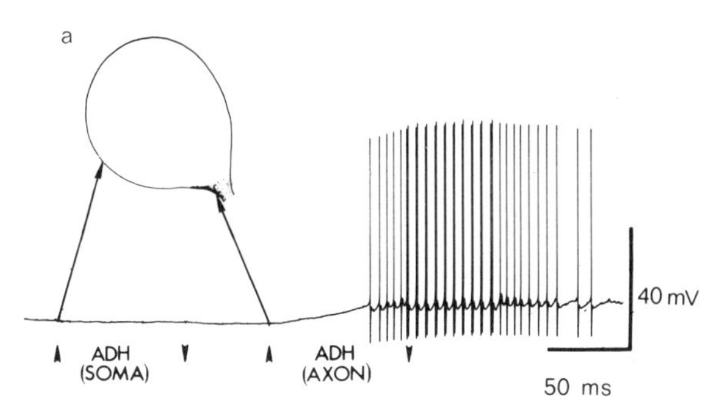

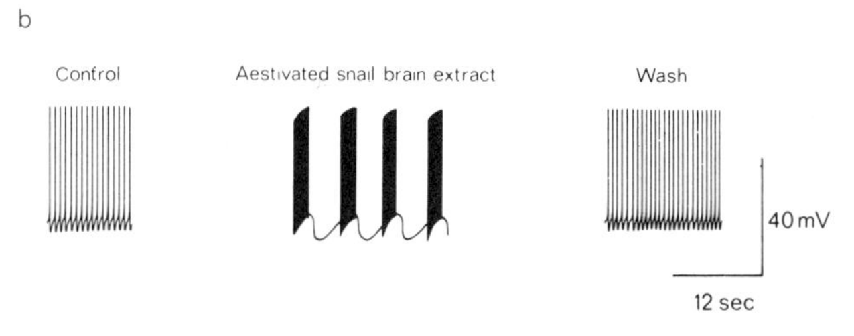

Figure 3.12 (*a*) Effects of lysine-vasopressin on the membrane properties of a neuroendocrine cell of a snail *Otala lactea*. Addition of vasopressin (ADH) to the soma has no effect, addition to the axon induces a bursting pacemaker potential which fades out after the peptide is washed off (from Barker *et al.*, 1975, ref. 40). (*b*). Effects of snail brain extracts on the bursting pacemaker potential (BPP) activity of a neuroendocrine cell of the snail *Otala lactea*. Addition of the extract of an aestivated snail induced BPP activity in a dormant snail. Washing for more than 1 h with extract-free saline restored the activity to approximately its control levels (from Ifshin *et al.*, 1975, ref. 29).

bursts which, as we have seen, would tend to increase the release of neurosecretory material from the cell terminals. It was also found that such steroid hormones as aldosterone and hydrocortisone at concentrations close to 10^{-6} M would have the opposite effect, leading to a decrease in hormone release.

That such a system of controls might operate *in vivo* gained a good deal of support from the subsequent finding that a peptide naturally occurring in the brain of the snail *Otala* would produce similar long-term changes in the sensitive neurosecretory cells of the animal[29] (figure 3.12). Interestingly, in view of the idea that bursting cells might be driven by changes in intracellular calcium levels generated by a feedback loop involving cyclic AMP (p. 43), it has been found that the peptide causes changes in the intracellular levels of cyclic AMP.

Such findings are most important because they suggest that neuro-secretory cells might be sensitive to the hormones they release, or other hormones in circulation which would act as neuromodulators. This could well be a highly significant element in the control of the concentration of hormone in circulation (see chapter 8, p. 155).

Electrical activity and hormone synthesis

Since it seems likely that increased electrical activity in a neurosecretory cell leads to an increase in the rate at which it liberates its neurosecretory material, the questions arise as to whether there might be increases in the rate of protein synthesis to replenish that which is lost and, if so, whether the electrical activity is itself the stimulus for the increase in protein synthesis.

Although there is evidence in squid axons and in the photoreceptor cells of the horseshoe crab that increased electrical activity is followed by more rapid incorporation of amino acids into proteins, there is as yet no such evidence for neurosecretory cells. Indeed, in one study of neurosecretory cells of the land snail *Otala*, it was found that amino acids were incorporated more rapidly into protein in electrically inactive cells than in spontaneously firing cells.[30] However, this work compared one cell type with another and they may well have differed in respects other than that of their electrical activity. The only suggestion of a correlation of electrical activity and protein synthesis is from a study of neurosecretory cell type in the brain of *Helix* thought to produce a hormone controlling hibernation. These cells synthesized protein rapidly at a time when the animals emerged from hibernation. Similar cells in another snail, *Otala*, undergo a grea increase in electrical activity at this time, so that it is probable that these cells simultaneously become active both electrically and in protein synthesis.[31,32] This does not mean to say, however, that the two activities must be causally linked, as some other stimulus might well be responsible for increasing both activities.

In chapter 2, we saw (p. 17) that hormone release does not lead to increases in the rate of axonal transport of neurosecretory granules. There was evidence, in the case of the mammalian hypothalamo-neurohypophysial tract, that much more hormone was synthesized than is ever normally released and that "old" neurosecretory granules are removed from the release sites and broken down. If this is general, then there would be no need for acceleration of protein synthesis after release, only for a reduction in the rate of granule destruction. It remains to be seen whether invertebrate systems are similar in this respect.

Depolarization and hormone release

The evidence so far set out in this chapter leads to the idea that hormone release follows the arrival at the axon terminals of action potentials which have travelled down the axon. We shall see in the next chapter how changes in the cytoplasmic calcium concentration play a central role in provoking hormone liberation. What electrical events occur at the axon ending that might lead to changes in the internal calcium level?

Because of the small size of most individual axon endings, it is not usually possible to make direct electrical recordings from them. However, in the sinus glands of crabs the neurosecretory axon terminals are large enough to permit intracellular recording.[7] As we have seen (p. 36), the action potential invades the ending causing depolarization at its surface. That this depolarization might be the trigger for changes in calcium level and subsequent hormone release seems very likely for the following reasons.

Neurosecretory axon terminals need to be in intimate contact with the blood, but in many cases the rest of the cells lie behind the blood-brain barrier (see p. 5). In many different animals, when their neurohaemal areas are treated with potassium-rich solutions there are rapid releases of hormones. Since potassium-rich solutions cause the depolarization of nearly all cell membranes, it is naturally thought that their effectiveness is due to a depolarizing action and, because of the speed of the response, it is supposed that the effect is on the exposed axon terminals rather than on the less-accessible pre-terminal axons. In addition, treatment with potassium-rich, but calcium-free, solutions almost always fails to cause liberation of significant amounts of hormone. This, of course, lends credence to the idea that depolarization has its effect by causing an influx of calcium ions (p. 66). Figures 3.13 and 3.14 show examples of hormone release stimulated by potassium-rich solutions and by acetylcholine, and show how release is affected by changes in potassium, sodium, and calcium concentrations and by time of exposure to these solutions.[34]

As figure 3.15 shows, the rate of hormone release declines during prolonged treatment with potassium-rich fluid. There could be several explanations of this. It might, for example, be that depolarization is not maintained. In most neurohaemal systems there is no direct way of testing this. However, it has been shown in the adrenal medulla that depolarization does not fail during prolonged exposure to potassium-rich solutions. If a

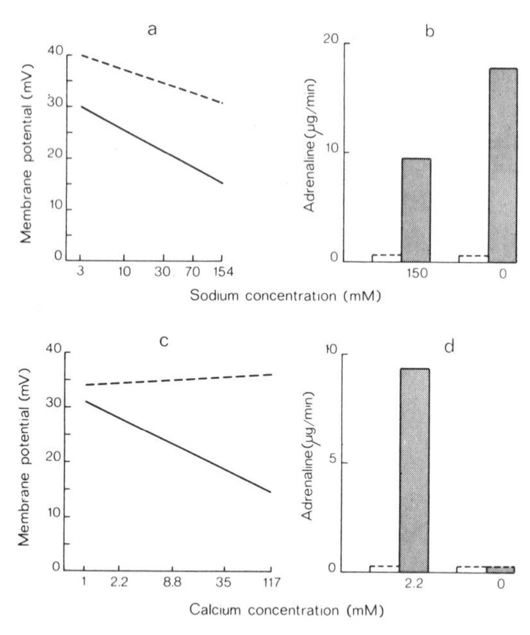

Figure 3.13 (*a*) Relation between external sodium concentration, resting potential (dashed line) and depolarization induced by acetylcholine (full line) in gerbil chromaffin cells *in vitro* (after Douglas *et al.*, 1967, ref. 37). (*b*) Relation between sodium concentration and adrenaline release induced by acetylcholine. White columns: control medium. Black columns: in presence of acetylcholine (after Douglas & Rubin, 1963, ref. 38). (*c*) Relation between external calcium concentration, resting potential (dashed line) and depolarization induced by acetylcholine (full line) in gerbil chromaffin cells *in vitro* (after Douglas *et al.*, 1967, ref. 37). (*d*) Relation between calcium concentration and adrenaline release induced by acetylcholine. White columns: control medium. Black columns: in presence of acetylcholine (after Douglas & Rubin, 1963, ref. 38).

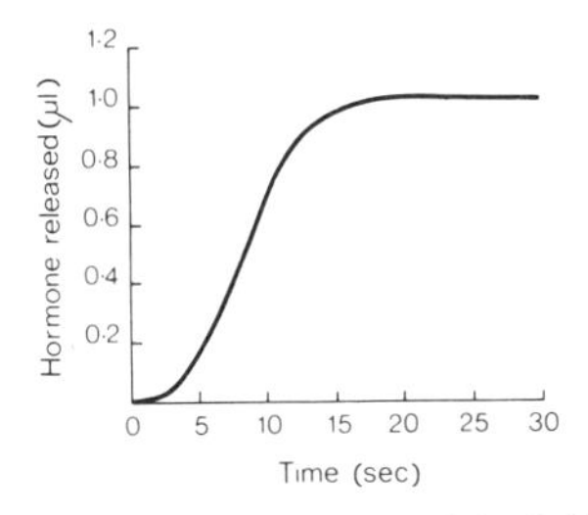

Figure 3.14 The quantities of diuretic hormone released (as fluid secretion) from release sites in *Rhodnius* in response to exposures of different lengths of time to a solution containing 70 mM K (after Maddrell & Gee, 1974, ref. 39).

potential-sensitive fluorescent dye is allowed to penetrate into the cell membrane, its behaviour shows that depolarization is maintained for as long as the cells are exposed to solutions rich in potassium.[33] If we accept that the same is true in conventional neurosecretory systems, then the steady failure of hormone release during treatment with potassium-rich solutions is not a failure of depolarization.

Another possibility is that the supply of hormone-containing granules (see pp. 21, 23) in the axon terminals might be exhausted. However as figure 3.15 shows, if a neurohaemal system is bathed first for an extended

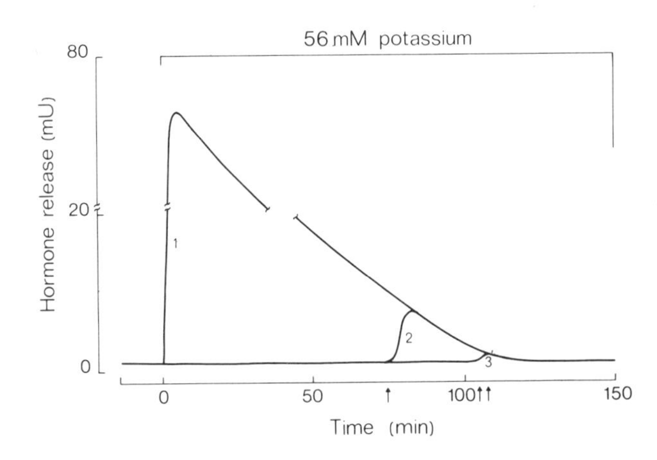

Figure 3.15 Effects of prolonged depolarization of the rat neurohypophysis on hormone release. Curve 1: Depolarization induces a large release of hormone which is not maintained during prolonged stimulation. Curve 2: The neurohypophysis depolarized in the absence of extracellular calcium. Later addition of calcium (single arrow) induces hormone release. Curve 3: As 2, but calcium was added later still (double arrow) (after Nordmann, 1976, ref. 35).

period in potassium-rich but calcium-free solution (in which no hormone is released) and then in a solution having the same concentration of potassium but now also containing calcium, only small amounts of hormone are released by exposure to the latter solution. Since this hormone release is similar in extent and time course to hormone release during the same late stage of potassium depolarization in a calcium-containing solution, it seems that hormone depletion of the axon terminal plays no part at all in the steady failure of hormone release in these short-term experiments.

Since calcium ions are essential to hormone release,[34] it might be that the rate at which calcium ions enter declines during prolonged depolarization, and it is this which reduces the rate of hormone release.[35] At the moment this looks to be the most likely explanation. Measurements of the uptake of ^{45}Ca by the rat neurohypophysis during potassium-depolarization have suggested that calcium uptake does indeed decline and it falls with a time course similar to that of the failure of hormone release.

Although the cells of the adrenal medulla are not typical of neurosecretory systems in that hormone release takes place at the cell body (there being no axons), the large size of the cells allows experiments to be done which so far have not been possible on more structurally typical neurosecretory systems. For example, adrenaline release is brought about *in vivo* by acetylcholine released at sympathetic synapses. Membrane potential recordings made during *in vitro* stimulation with acetylcholine show a strong depolarization associated with hormone release[36] (figure 3.16). Furthermore, recordings of the membrane potential in the presence

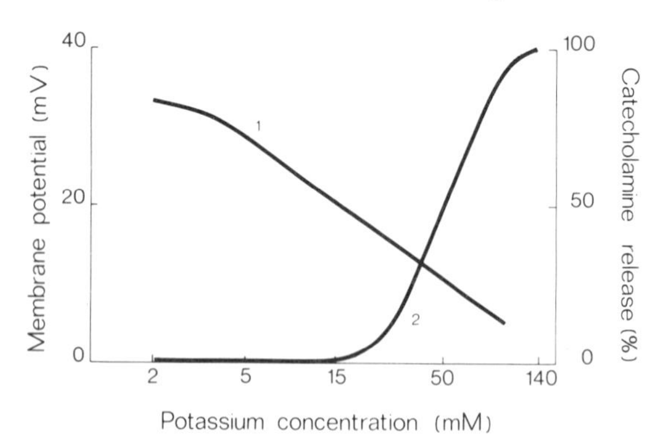

Figure 3.16 Relation between potassium concentration of the medium, membrane potential (1) and catecholamine release (2) in chromaffin cells (1 after Douglas *et al.*, 1971, ref. 37;2 courtesy of T. J. Rink).

or absence of acetylcholine show that depolarization depends on the presence of either sodium or calcium ions in the bathing solution (figure 3.13). To have such an effect these ions must cross the cell membrane; since hormone release only occurs in calcium-containing solutions, even in those

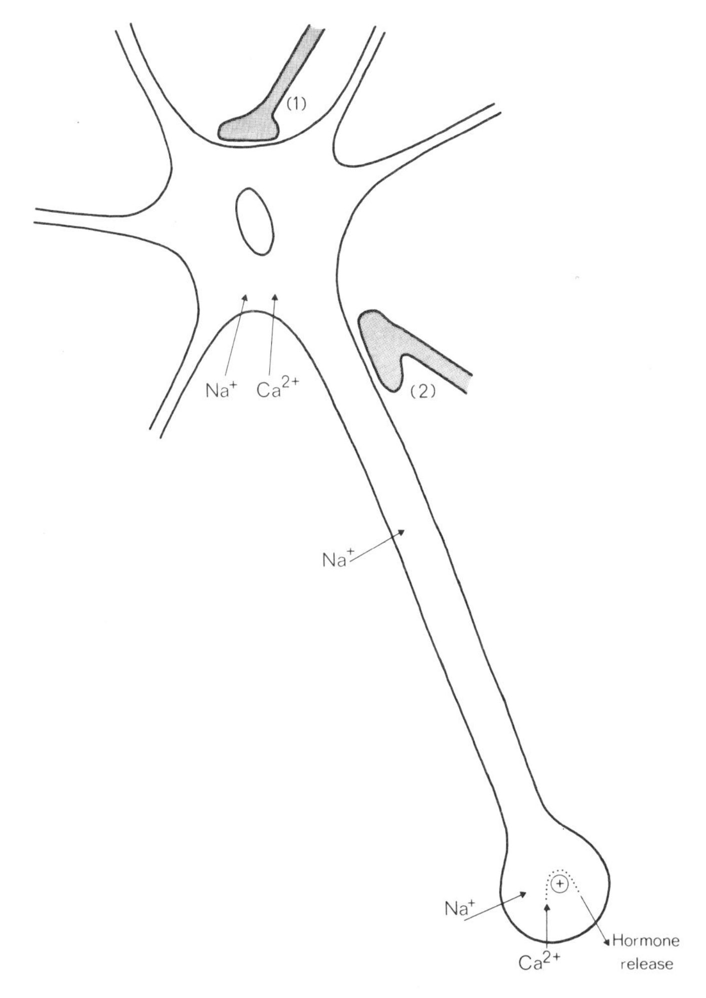

Figure 3.17 A summary of the electrical events occurring in neurosecretory cells showing the ions responsible for carrying the inward current during electrical activity. Stimulation of the cell occurs either at the cell body (synapse 1) or, in invertebrates, in the neuropile (2). An increase in the permeability to Na and Ca gives rise to an action potential which, relying on Na entry, is propagated along the axon. The resulting depolarization of the nerve terminal promotes Ca entry, which in turn induces hormone release.

which lack sodium, it follows that hormone release depends solely on calcium influx. Clearly depolarization of the nerve terminal is not in itself sufficient to cause hormone release, as depolarization in calcium-free solutions is ineffective.

As mentioned earlier (p. 36), it is possible to record changes in membrane potential at the neurosecretory axon endings of crab sinus glands. In solutions containing only half the normal sodium concentration or containing tetrodotoxin (TTX) at 10^{-7} M (this poison blocks sodium entry during nerve impulses), there are no membrane potential changes following stimulation of the axon. However, this turns out only to be a failure of axonal conduction, for stimulation of the endings themselves causes regenerative responses in the membrane potential, even in solutions containing only 5% of the normal sodium level or 10^{-6} M TTX. The evidence suggests that this is because part of the inward current during the action potential at the terminal is carried by calcium ions. These results are consistent with those from work on the adrenal medulla. Together they provide support for the idea that the release of neurohormones at neurosecretory axon endings is brought about by an influx of calcium during depolarization of the membrane following the arrival of axonal action potentials.

Figure 3.17 summarizes the electrical events believed to occur in neurosecretory cell bodies, their axons and their terminals, and emphasizes those features which are significant for hormone release.

REFERENCES

1. Kandel, E. R. (1964) "Electrical properties of hypothalamic neuroendocrine cells," *J. Gen. Physiol.* **47,** 691–717.
2. Yamashita, H., Koizumi, K. & Brooks, C. McC. (1970) "Electrophysiological studies of neurosecretory cells in the cat hypothalamus," *Br. Res.* **20,** 462–466.
3. Cross, B. A. & Green, J. D. (1959) "Activity of single neurones in the hypothalamus: effect of osmotic and other stimuli," *J. Physiol.* **148,** 544–569.
4. Yagi, K., Azuma, T. & Matsuda, K. (1966) "Neurosecretory cell: capable of conducting impulse in rats," *Science*, **154,** 788–779.
5. Standen, N. B. (1975) "Calcium and sodium ions as charge carriers in the action potential of an identified snail neurone," *J. Physiol.* **249,** 241–252.
6. Geduldig, D. & Junge, O. (1968) "Sodium and calcium components of action potentials in the *Aplysia* giant neurons," *J. Physiol.* **199,** 347–365.
7. Cooke, I. M. (1977) "Electrical activity of neurosecretory terminals and control of peptide hormone release," pp. 345–374 in *Peptides in Neurobiology* (ed. H. Gainer), New York, Plenum Press.
8. Dunn, F. L., Brennan, T. J., Nelson, A. E. & Robertson, G. L. (1973) "The role of blood osmolality and volume in regulating vasopressin secretion in the rat," *J. Clin. Invest.* **52,** 3212–3219.

9. Koizumi, K. & Yamashita, H. (1972) "Studies of antidromically identified neurosecretory cells of the hypothalamus by intracellular and extracellular recordings," *J. Physiol.* **221**, 683–705.
10. Hayward, J. N. & Vincent, J. D. (1970) "Osmosensitive single neurones in the hypothalamus of unanaesthetized monkeys," *J. Physiol.* **210**, 947–972.
11. Dyball, R. E. J. & Poutney, P. S. (1973) "Discharge patterns of supraoptic and paraventricular neurones in rats given a 2% NaCl solution instead of drinking water," *J. Endocr.* **56**, 91–98.
12. Arnauld, E., Vincent, J. D. & Dreifuss, J. J. (1974) "Firing patterns of hypothalamic supraoptic neurons during water deprivation in monkeys," *Science* **185**, 535–537.
13. Dyball, R. E. J. (1974) "Single unit activity in the hypothalamo-neurohypophysial system of Brattleboro rats," *J. Endocr.* **60**, 135–143.
14. Poulain, D. A., Wakerley, J. B. & Dyball, R. E. J. (1977) "Electrophysiological differentiation of oxytocin and vasopressin-secreting cells," *Proc. R. Soc. Lond. Ser. B.* **196**, 367–384.
15. Gillary, H. L. & Kennedy, D. (1969) "Neuromuscular effects of impulse pattern in a crustacean motoneuron," *J. Neurophysiol.* **32**, 607–612.
16. Nordmann, J. J. & Dreifuss, J. J. (1972) "Hormone release evoked by electrical stimulation of rat neurohypophyses in the absence of action potentials," *Br. Res.* **45**, 604–607
17. Alving, B. O. (1968) "Spontaneous activity in isolated somata of *Aplysia* pacemaker neurons," *J. Gen. Physiol.* **51**, 28–45.
18. Gainer, H. (1972) "Electrophysiological behaviour of an endogenously active neurosecretory cell," *Br. Res.* **39**, 403–418.
19. Stinnakre, J. & Tauc, L. (1969) "Central neuronal response to the activation of osmoreceptors in the osphradium of *Aplysia*," *J. exp. Biol.* **51**, 347–361.
20. Meech, R. W. (1972) "Intracellular calcium injection causes increased potassium conductance in *Aplysia* neurones," *Comp. Biochem. Physiol.* **42A**, 493–499.
21. Meech, R. W. & Standen, N. B. (1975) "Potassium activation in *Helix aspersa* neurones under voltage clamp: a component mediated by calcium influx," *J. Physiol.* **249**, 211–239.
22. Gorman, A. L. F. & Thomas, M. V. (1978) "Changes in the intracellular concentration of free calcium ions in a pace-maker neurone, measured with the metallochromic indicator dye Arsenazo III," *J. Physiol.* **275**, 357–376.
23. Rapp, P. E. & Berridge, M. J. (1977) "Oscillations in calcium-cyclic AMP control loops form the basis of pacemaker activity and other high frequency biological rhythms," *J. theor. Biol.* **66**, 497–525.
24. Treistman, S. N. & Levitan, I. B. (1976) "Alteration of electrical activity in molluscan neurones by cyclic nucleotides and peptide factors," *Nature* **261**, 62–64.
25. Strumwasser, F. (1973) "Neural and humoral factors in the temporal organization of behaviour," *The Physiologist* **16**, 9–42.
26. Both, R., Finger, W. & Chaplain, R. A. (1976) "Model predictions of the ionic mechanisms underlying the beating and bursting pacemaker characteristics of Molluscan neurons," *Biol. Cybernetics.* **23**, 1–11.
27. Dreifuss, J. J., Nordmann, J. J. & Vincent, J. D. (1974) "Recurrent inhibition of supraoptic neurosecretory cells in homozygous Brattleboro rats, "*J. Physiol.* **237**, 25–27P.
28. Barker, J. L. & Gainer, H. (1974) "Peptide regulation of bursting pacemaker activity in a molluscan neurosecretory cell," *Science* **184**, 1371–1373.
29. Ifshin, M., Gainer, H. & Barker, J. L. (1975) "Peptide factor extracted from molluscan ganglia that modulates bursting pacemaker activity," *Nature* **254**, 72–74.
30. Loh, Y. P. & Gainer, H. (1975) "Low molecular weight specific proteins in identified molluscan neurons. I. Synthesis and storage," *Brain Res.* **92**, 181–192.
31. Loh, Y. P., Barker, J. L. & Gainer, H. (1976) "Neurosecretory cell protein metabolism in the land snail, *Otala lactea*," *J. Neurochem.* **26**, 25–30.
32. Gainer, H. (1972) "Effects of experimentally induced diapause on the electrophysiology and protein synthesis of identified molluscan neurons," *Br. Res.* **39**, 387–402.

33. Baker, P. F. & Rink, T. J. (1975) "Catecholamine release from bovine adrenal medulla in response to maintained depolarisation," *J. Physiol.* **253,** 593–620.
34. Douglas, W. W. (1968) "Stimulus-secretion coupling: the concept and clues from chromaffin and other cells," *Br. J. Pharmac.* **34,** 451–474.
35. Nordmann, J. J. (1976) "Evidence for calcium inactivation during hormone release in the rat neurohypophysis," *J. exp. Biol.* **65,** 669–683.
36. Douglas, W. W., Kanno, T. & Sampson, S. R. (1967) "Effects of acetylcholine and other medullary secretagogues and antagonists on the membrane potential of adrenal chromaffin cells: an analysis employing techniques of tissue culture," *J. Physiol.* **188,** 107–120.
37. Douglas, W. W., Kanno, T., and Sampson, S. R. (1967) "Influence of the ionic environment on the membrane potential of adrenal chromaffin cells and on the depolarizing effect of acetylcholine," *J. Physiol.* **191,** 107–121.
38. Douglas, W. W. & Rubin, R. P. (1963) "The mechanism of catecholamine release from the adrenal medulla and the role of calcium in stimulus-secretion coupling," *J. Physiol.* **167,** 388–410.
39. Maddrell, S. H. P. & Gee, J. D. (1974) "Potassium-induced release of the diuretic hormones of *Rhodnius prolixus* and *Glossina austeni*: Ca dependence, time course and localization of neurohaemal areas." *J. exp. Biol.* **61,** 155–171.
40. Barker, J. L., Ifshin, M. S. & Gainer, H. (1975) "Studies on bursting pacemaker potential activity in molluscan neurons. III. Effects of hormones," *Br. Res.* **84,** 501–513.

CHAPTER FOUR

RELEASE OF NEUROHORMONES

THE LAST CHAPTER SHOWED HOW HORMONE RELEASE FROM NEUROSECRETORY axon endings is triggered by action potentials which depolarize the endings. It also gave some evidence that release depends on the entry of calcium ions.

In the present chapter we shall see how hormone present in membrane-bound granules is liberated into the extracellular space. Then we shall look further at the role of calcium in the process. Finally we shall consider the fate of the membranes surrounding the granules which contain the neuro-hormones prior to their release.

Possible mechanisms of release of neurohormones

The many possible mechanisms of release can be divided into three broad categories (figure 4.1):

(*a*) Those in which the granule contents are first released into the cytoplasm of the axon endings, from which they pass by diffusion out into the extracellular fluid.
(*b*) Those in which the granules pass intact through the plasma membrane of the endings before liberating their contents.
(*c*) Those in which the granules contact the plasma membrane and release their contents, either through a low-resistance pathway into the extracellular fluid, or more directly, by a process of membrane fusion in which the granule membranes and that of the axon become continuous, so that the granule contents effectively lie outside the cell. This process is known as exocytosis.

At an early stage it seemed that the evidence supported a mechanism of type (*a*), i.e. one in which vesicle contents were liberated intracellularly into the cytoplasm. For example, early experiments in which the cellular contents of the rat neurohypophysis were fractionated, initially showed that more than 50% of the hormone content was to be found in the cytoplasm. Furthermore, there was ultrastructural evidence that in several neurohormone-releasing tissues, many of the neurosecretory granules in the axon endings appeared pale, with the surrounding membrane incomplete, and often a proportion of the granules were completely electron-transparent as if their contents had been lost. Against this

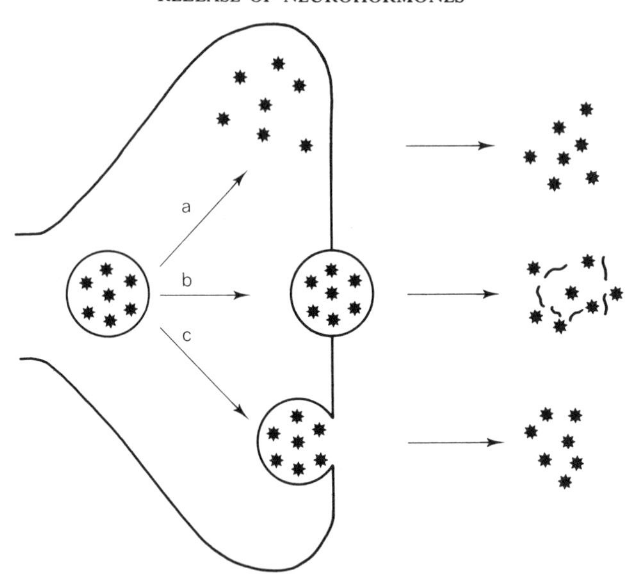

Figure 4.1 Possible mechanisms for release of neurohormones: (*a*) via the cytoplasm through the cell membrane; (*b*) the entire granule is extruded and its membrane is destroyed in the extracellular space; (*c*) via exocytosis which involves fusion and fission between the granule membrane and the plasma membrane.

evidence, however, has to be set the following strong objections on theoretical grounds. First of all, compounds as physiologically active as hormones appearing in the cytoplasm would be expected to have adverse effects on the cell. Secondly, the hormone would have to pass out through the cell membrane. Unless the membrane was very permeable, when there would be a steady drain of other substances of similar size and charge from the cytoplasm, this would require some process of facilitated passage to speed it up. Thirdly, the concentrated hormone from the granules would become diluted in the volume of the axon endings, and this would slow its appearance in the extracellular fluid. Fourthly it seems likely that the hormone might be damaged by the activity of cytoplasmic enzymes—although one can argue that hormones might well have been evolved with a resistance to this.

More recent work has now greatly weakened the evidence supporting a mechanism of type (*a*). As cell fractionation techniques have steadily improved, more and more of the hormone content of, for example, the rat neurohypophysis, turns out to be contained within the granule fraction. Furthermore, the pale granules seen in the early studies of these tissues turn out to be due to an artefact of fixation procedures carried out at alkaline pHs[1] (see figure 4.2). Under acid conditions of fixation, however, electron-transparent granules are still found. Moreover, they occur in numbers which are correlated with the rate at which the tissue was releasing its neurohormone content in the period before fixation. However, it turns out that these empty vesicles are not involved in the actual process of release, but are products of a membrane retrieval mechanism (p. 73). On all these grounds there seems little reason at present for supposing that neurohormones are released from a cytoplasmic pool.

If there is little evidence for release mechanism of type (*a*), there is even

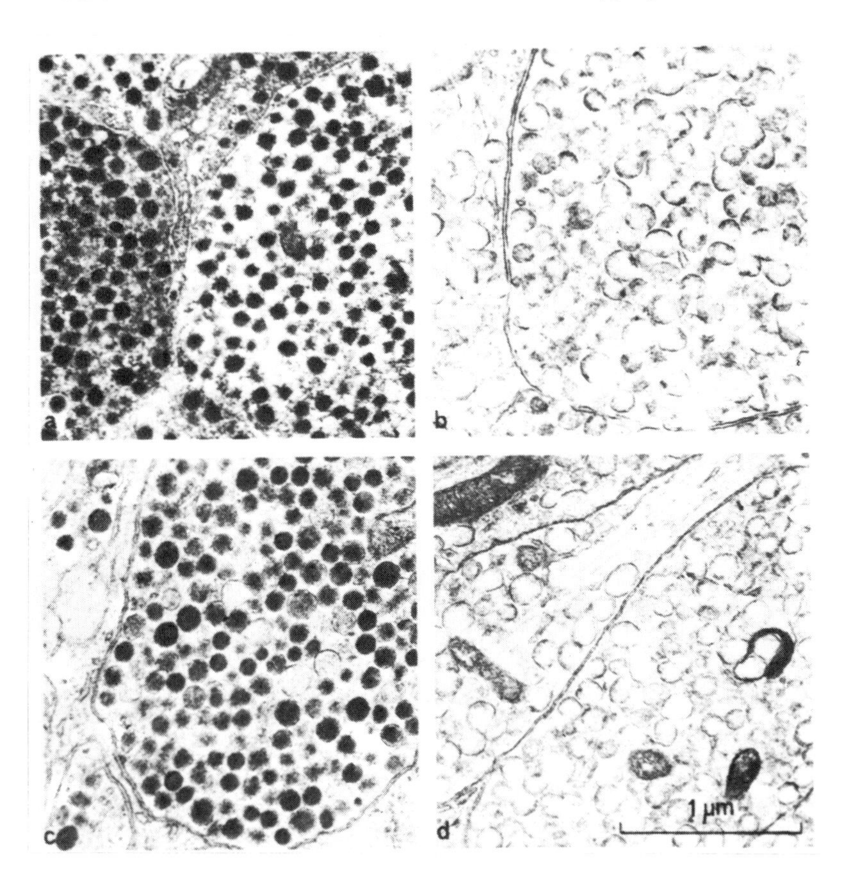

Figure 4.2 Effect of pH of the fixative on electron density of neurosecretory granules in the rat neural lobe. In (*a*), fixation was at pH 5·0 for 2 hours, in (*b*) at pH 8·0 for 2 hours, in (*c*) at pH 7·3 for 20 minutes, and in (*d*) at pH 7·3 for 24 hours (after Morris and Cannata, 1973, ref. 1).

less evidence for a mechanism of type (*b*) in which intact neurosecretory granules would be extruded from the axon endings before releasing their contents. If such a process occurred, one would expect to find ultrastructural evidence for it, but no intact granules have been seen extracellularly in systems known to be releasing neurohormones. Indeed, it has been found that if noradrenaline-containing vesicles from the adrenal medulla are injected into the blood, they produce none of the effects that would be expected from their content of noradrenaline. So it seems likely that the extrusion of intact neurosecretory granules would not in itself lead to the release of their contents; it would therefore be necessary, if such a mechanism is to occur, that the vesicle membrane be altered in some way, perhaps enzymatically, either as it crosses the cell membrane or in the extracellular environment. No evidence for this has so far been found. Furthermore, during medullary stimulation leading to release of adrenaline and noradrenaline, there is no release of phospholipid and cholesterol,[2] although they are both major components of the membrane of the chromaffin granules.

There has recently appeared conclusive evidence that release does not

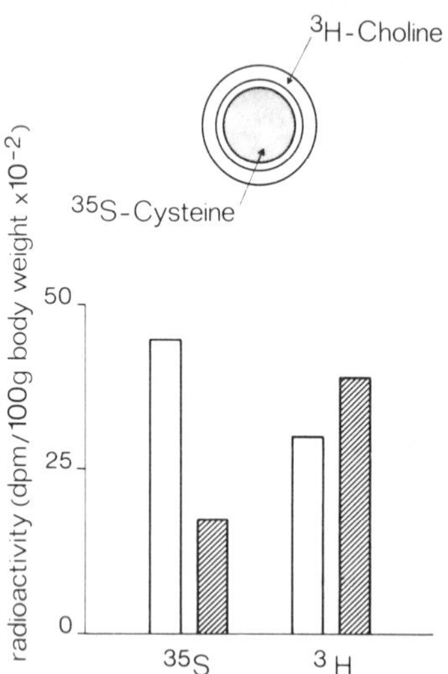

Figure 4.3 An experiment to show that only the content of a neurosecretory granule is released; the granule membrane remains in the gland. The granule membrane was labelled by injecting ^{3}H-choline into the third ventricle, whereas the granule content was labelled with a similar injection of ^{35}S-cysteine. Hormone release was induced by allowing the rats to drink 2% NaCl solution. In each case the unshaded bars indicate the radioactivity in the pituitary before release and the cross-hatched bars indicate the radioactivity remaining in the pituitary after release (modified from Swann & Pickering, 1976, ref 3).

occur by a mechanism of type (*b*). ^{35}S-cysteine and ^{3}H-choline were injected into the third ventricle of the brain in rats. These labels appeared in neurosecretory granules arriving at the pituitary.[3] ^{35}S-cysteine appeared in the granule contents and ^{3}H-choline was incorporated into the granule membrane. When hormone release was stimulated, only ^{35}S appeared in the fluid surrounding the neurohypophysis[3] (figure 4.3).

In contrast, evidence for a mechanism in which neurohormones are released by exocytosis has steadily accumulated in the last ten years.[4,5] Exocytosis involves the fusion of the granule membrane with that of the axon ending (figure 4.1) so that the neurohormonal content can diffuse freely away in the extracellular fluid.

Ultrastructural evidence for exocytosis

At the moment at which a neurosecretory granule fuses with the plasma membrane, the joined membranes show a characteristic exocytotic profile, the so-called omega figure (figure 4.1). In a variety of tissues known to be releasing hormone either naturally or *in vitro* following extrinsic stimulation, there are large increases in the numbers of such profiles visible. Such evidence has been found, for example, in the release of adrenaline from the adrenal glands of the hamster,[6] in the neurohypophysis of the platyfish,[7] the hamster and the rat,[8] and amongst such invertebrate neurohaemal organs and systems as the corpora cardiaca of flies (*Calliphora*)[9] and of *Rhodnius* (figure 4.4), the perisympathetic organs of the stick insect (*Carausius*)[10] and the hormone-releasing surfaces of the abdominal nerves of the bloodsucking insect *Rhodnius*[11] (details of these very different hormone-releasing systems are given in the next chapter, p. 79).

In other systems, however, exocytotic profiles are encountered so rarely as to cast real doubt on the importance of this process in release. However, this does not mean that exocytosis is not the dominant method of release. If the process of exocytosis is swift, an omega figure would persist only briefly, and the chance of fixing it at any one moment would be correspondingly reduced. Calculations based on the known rates of release of hormone from the rat neurohypophysis are particularly telling in this context. Under maximal stimulation, approximately 10^8 neurosecretory granules are released per minute from a population of 4×10^7 neurosecretory axon endings; i.e. each ending releases only 1–2 granules per minute. If the exocytotic process is complete in, say, 5 ms, only 1 ending in 8000 will actually be in the process of releasing a neurosecretory granule at any one moment. Add to this the slim chance of this granule appearing in

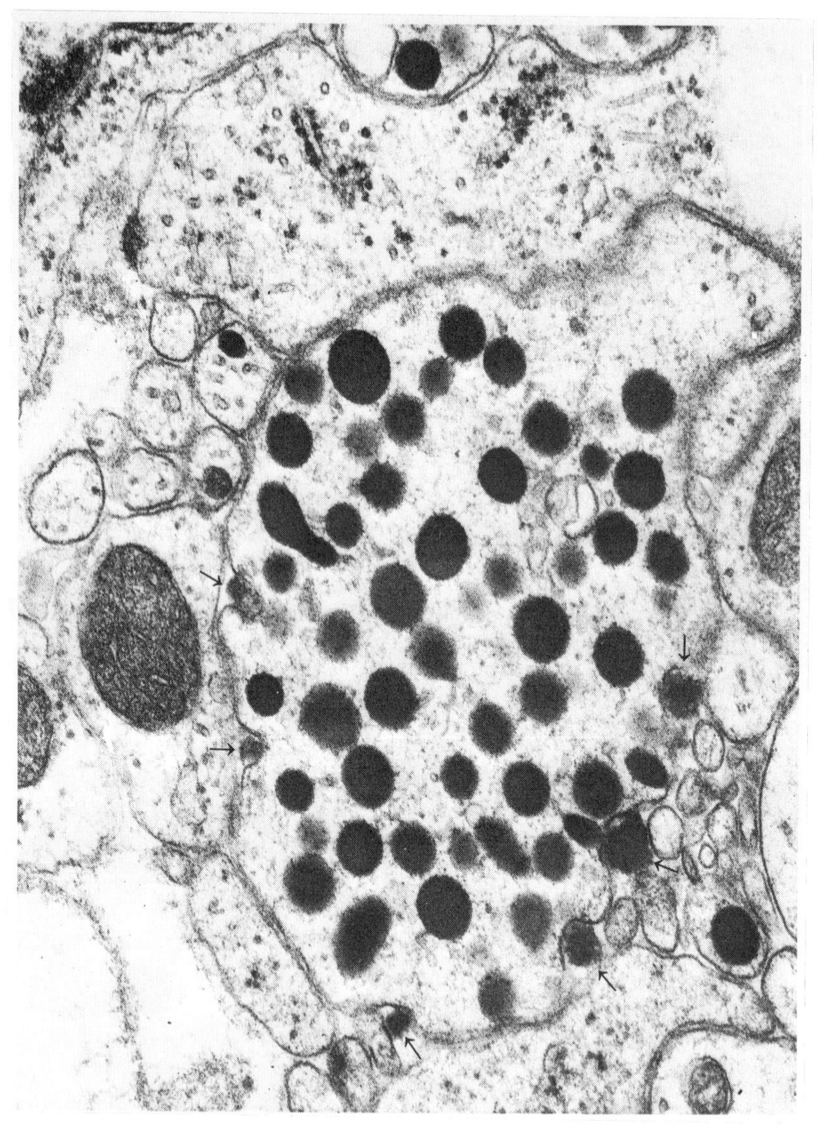

Figure 4.4 Exocytosis in the corpus cardiacum of *Rhodnius prolixus*. Note the extruded granule contents (indicated by arrows). (Micrograph courtesy of G. P. Morris and C. G. H. Steel. × 47 500)

the plane of the very thin sections used in electron microscopy, and one can appreciate just how few omega profiles are to be expected in a single section, even of neurohaemal tissue releasing hormone at a high rate.

When some tissues are fixed, vesicles which lie just beneath the cell surface come to be fused to the cell membrane in an artifact resembling exocytosis.[12] They do not do this if the tissue is not fixed but rapidly frozen and examined by the "freeze fracture" technique (figure 4.5). This technique involves cutting the tissue in a deeply frozen state. The tissue fractures rather than cuts cleanly. Still in the frozen state, the fractured surface of the specimen is coated with a heavy metal sublimed on to it in a vacuum. The tissue is then washed away and the metal replica of the surface is examined by conventional electron microscopy.

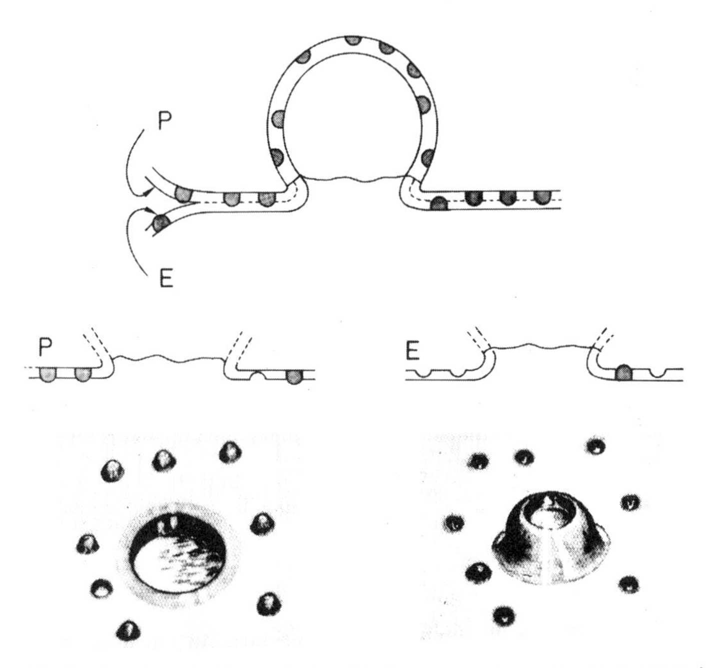

Figure 4.5 To show how the "freeze fracture" technique reveals membrane structure, in this case of an exocytotic profile. Membranes tend to fracture between their two component surfaces—as indicated by the dotted line in the upper diagram. The face of the inner cell membrane thus revealed is termed the P face; that of the outer cell membrane is the E face. The appearances of these two faces are shown in the lower part of the figure, the P face on the left and the E face on the right. The uppermost drawing in each case shows the face in cross-section, while the lower drawings represent the appearance of the metal replicas of the two faces that are examined by electron microscopy. The shaded objects in the membranes represent membrane proteins.

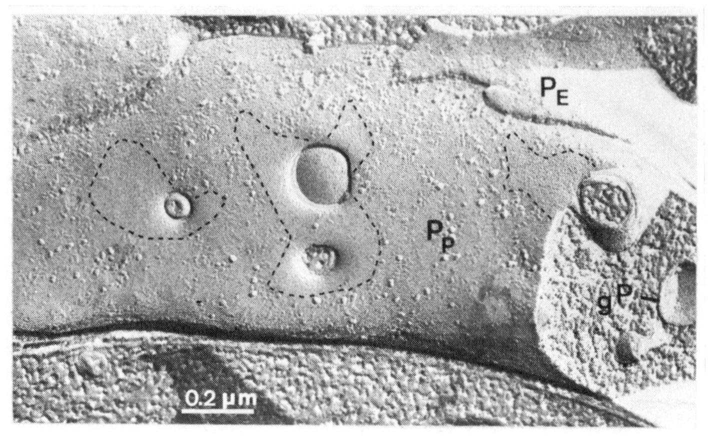

Figure 4.6 Electron micrograph of changes indicative of exocytosis in membranes of freeze-fractured neurosecretory axons of the dormouse neurohypophysis. When the fracture passes through the interior of the plasma membrane, two membrane faces are exposed; an inner P face (Pp), adjacent to the cytoplasm, and an outer E face (P_E), adjacent to the extracellular space. The fracture is here seen to have passed not only through the plasmalemma, but also through the cytoplasm of the same cell (gP, secretory granule membrane P-face). The plasmalemma P-face shows a much higher density of randomly distributed intramembranous particles than the E face. However, the number of particles markedly diminishes in areas adjacent to the exocytotic depressions (dotted lines). On passing into the cell interior, the fracture reveals that the *en face* depressions are invaginations into the cell interior, filled with granule cores in the process of extrusion into the extracellular space. The granule cores appear smooth-surfaced when viewed *en face* or finely granular when cross-fractured. (Micrograph courtesy of D. Theodosis. × 50 000; compare with fig. 4.5.)

This technique has been applied to neurosecretory tissue and has shown that exocytosis still occurs (figure 4.6). This strengthens the support for the belief that exocytosis is a real phenomenon involved in the release of neurohormones.

Biochemical evidence for exocytosis

As figure 4.1 and 4.5 indicate, if exocytosis is the mechanism of release, then it follows that:

(1) All substances present in the granules will be released.
(2) Substances occurring in the cytoplasm will not be released.

Perhaps the most convincing evidence for exocytosis comes from biochemical studies of the substances that appear, during stimulation, in the extracellular fluid in contact with neurosecretory endings. Preparations of neurosecretory granules show them to contain not only the active principle, but other compounds as well. For example, vesicles isolated from the adrenal medulla contain not only noradrenaline but also ATP and the soluble enzyme, dopamine-β-hydroxylase. Analysis of fluid which has been perfused through stimulated adrenal glands shows that it contains noradrenaline, ATP and dopamine-β-hydroxylase;[13] further, it is possible to measure the concentration of these substances and they occur in the same relative amounts as in the isolated granules (figure 4.7).

Granules isolated from the rat neurohypophysis contain neurophysins as well as neurohormones (p. 10). After stimulation of release of oxytocin and/or vasopressin, either *in vivo*[14,15] or *in vitro*[16,17] neurophysins are always found to occur in the extracellular fluid in amounts corresponding to the extent of hormone release[18] (figure 4.7).

One might perhaps be tempted to argue that the appearance of these other substances in the extracellular fluid during hormone release might follow from a general increase in the permeability of the plasma membrane. This is ruled out because substances of such disparate molecular size as ATP and dopamine-β-hydroxylase, for example, would be very unlikely to diffuse across the cell membrane at the same rate. Conclusive evidence comes from the finding that compounds of similar size, known to occur free in the cytoplasm, never appear in the extracellular fluid. For example, in the rat neurohypophysis, there is no evidence that trace concentrations of such cytoplasmic markers as adenylate kinase[18] and LDH[19] (lactate dehydrogenase) appear in the extracellular fluid when neurohormones are being liberated. Figure 4.8 summarizes this point.

Alternative forms of exocytosis

As we can see, the evidence for release of hormones by exocytosis is strong. It is generally believed that exocytosis involves the complete fusion of the membrane of single granules with the plasmalemma (external cell membrane). However, it is conceivable that exocytosis might effectively be achieved in a different way. The essence of exocytosis is the release, after fusion with the cell membrane, of the entire soluble contents of granules without release of material from the general cytoplasm or from the granule membrane. This might occur by fusion of single granules with the plasmalemma, but it could also be done by the fusion of several granules with only one actually fusing with the cell membrane. Some instances of what has been termed "compound exocytosis"[8] have been seen in neurosecretory nerve endings of the crab sinus gland, though none have been found in the neurohypophysis or in chromaffin cells. We have already

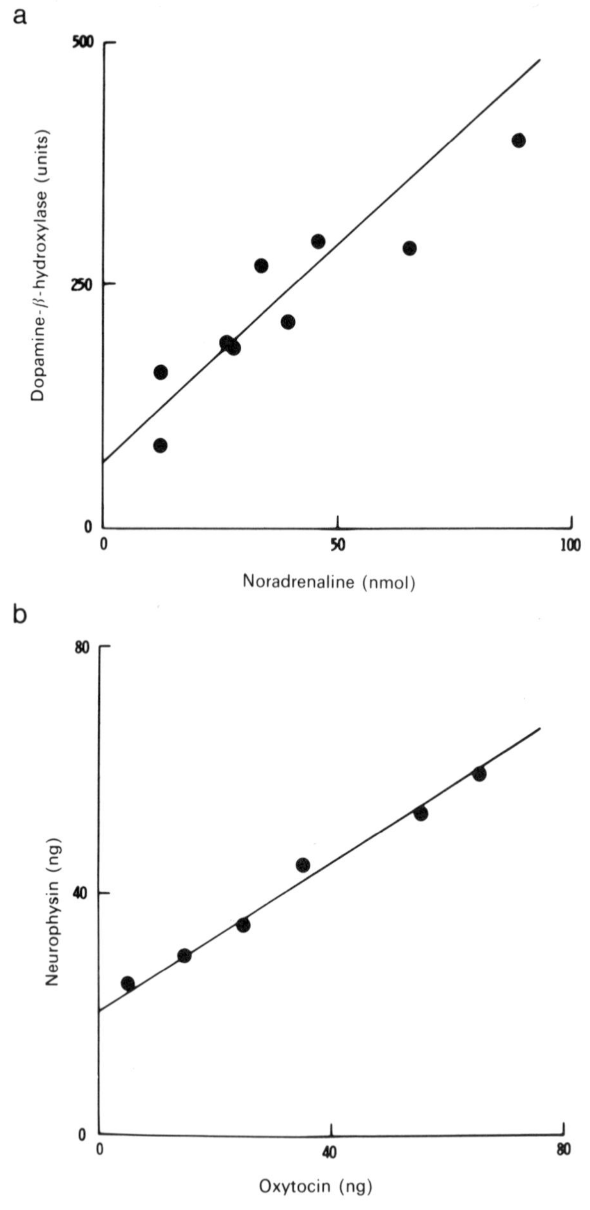

Figure 4.7 Release of noradrenaline and dopamine-β-hydroxylase from the adrenal medulla (*a*) and oxytocin and neurophysin from the neurohypophysis (*b*) ((*a*) modified from Smith A. D., ref. 51 and (*b*) modified from Nordmann *et al.*, ref. 16).

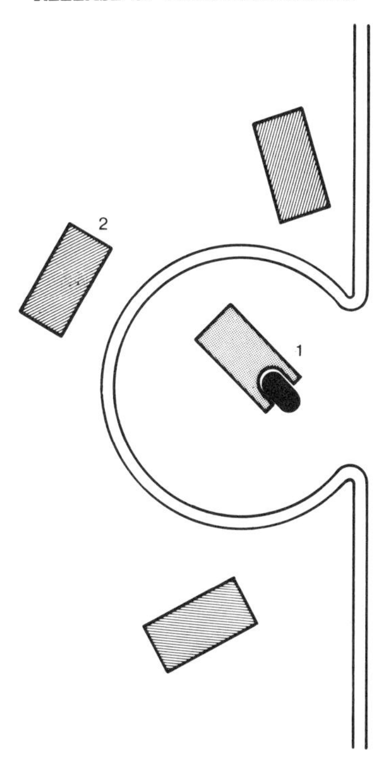

Figure 4.8 The biochemical conditions which need to be satisfied to demonstrate exocytosis (1) both large and small molecules found in the granule must be released, whereas (2) a cytoplasmic molecule of the same molecular weight as the largest found in the granule is not secreted.

seen at the end of chapter 3 that hormone release is calcium-dependent, so it is of interest to note that when neurosecretory granules are incubated *in vitro* they can be made to fuse with each other when the calcium concentration of the medium is raised.[20]

Finally it is worth mentioning that exocytosis could conceivably occur without membrane fusion at all. If the granule membrane were able to attach to the cell membrane forming a "gap junction" with it, even large molecules could readily escape through the apposed membranes. It remains to be seen whether such a system actually occurs.

Calcium and release of hormones

In other systems which are known to liberate physiologically active compounds from intracellular membrane-bound vesicles, such as the nerve endings at neuromuscular junctions, it is found that release depends on the presence of calcium ions in the bathing solution in contact with the sites of release. It turns out that the liberation of neurohormones is similarly calcium-dependent. The first evidence of this was from work on rats on the release of catecholamines, such as adrenaline, from the adrenal medulla[21] and on the release of neurohormones from the neurohypophysis.[22] The essential role of calcium in hormone release in these two systems was perhaps most convincingly shown in experiments where the organs were stimulated while being bathed in calcium-containing solution lacking sodium, potassium, and chloride ions; under these conditions electrical stimulation (neurohypophysis) or addition of acetylcholine (adrenal medulla) gave rise to very effective release of hormone. Since that time, a similar calcium dependence has been shown for neurohormone release systems in a variety of animals, both vertebrate and invertebrate (see refs. 23 & 24 for example). Because such different animals as insects, crustaceans and mammals, for example, all possess calcium-dependent release systems, it seems likely that the process was evolved at an early stage. We must now consider how calcium is involved in the process of exocytosis which, as we have seen, leads directly to the liberation of the contents of neurosecretory granules.

First of all it has to be pointed out that the picture is clouded somewhat by the finding that some exocytosis at least can go on in the absence of calcium ions from the bathing solution. For example, measurable amounts of oxytocin and vasopressin are slowly released from isolated unstimulated rat neurohypophyses bathed in calcium-free solutions.[18] This basal level of release seems still to be attributable to exocytosis, because neurophysins are released but not cytoplasmic proteins.

A further indication that extracellular calcium might not be essential comes from experiments where it was found that hormone release could be stimulated by treatment with particular calcium ionophores (compounds which facilitate the diffusion of specific substances across lipid membranes) made up in solutions lacking calcium.[25,26]

In resolving this apparent paradox, one has to remember that although the levels of calcium in the cytoplasm are generally only of the order of 10^{-7}M, there exist intracellular reservoirs, such as mitochondria, which contain much higher calcium concentrations. If it is the cytoplasmic level of the calcium which determines the rate of hormone release, then all these findings become intelligible. An increase in cytoplasmic level might be achieved either by a release of calcium from internal sources or by an increased rate of entry from the external fluid. If so, then slow release at a basal level, which does not require extracellular calcium, might reflect the cytoplasmic concentration of calcium derived from internal sources. Faster release, stimulated electrically for example, which is strongly dependent on external calcium, might involve an increase in the cytoplasmic level from calcium ions entering the cell through the plasma membrane.

Although there is as yet no direct proof of this in neurohormone-releasing systems, it has been shown that an increased cytoplasmic concentration of calcium ions leads to an increase in neurotransmitter release in the squid giant axon.[27] Nevertheless, in some neurosecretory systems it has at least been shown that, during hormone release, uptake of radioactive calcium ions into the system is much increased.[28] Furthermore, it seems very probable that the hormone release evoked by certain ionophores (referred to earlier) results from the action of these ionophores on internal calcium reservoirs. They are known, for example, to cause massive releases of calcium from suspensions of mitochondria from brains of guinea pigs. Hormone release *in vivo* follows the passage of nerve impulses along the axons to their endings. As we have seen (p. 36), the arrival of the action potential at the axon ending causes a temporary depolarization of the plasma membrane. As we saw on p. 47, one of the most effective ways of stimulating hormone release *in vitro* is to treat the neurohaemal areas with potassium-rich solutions which are known to depolarize cell membranes. These treatments not only cause release of hormone but also stimulate calcium uptake. Presumably depolarization of the membrane results in an increase in calcium permeability. Indeed, treatment with such agents as cobalt ions, manganese ions, or the drug D600, which are known to reduce calcium permeability of excitable membranes[29] greatly reduce hormone release, whether evoked by K-rich solutions or by electrical stimulation.[30] Treatment of the neurohypophysis with K-rich solutions causes large-scale releases of hormone, but still larger amounts are released if the K-rich solutions lack sodium ions.[22] One explanation that has received wide support is that sodium and calcium ions compete for entry and, in the absence of sodium ions, calcium ions can cross the membrane more freely and so promote a more extensive release of hormone.

The action of calcium in eliciting hormone release

All the evidence, then, is consistent with the idea that hormone release by exocytosis is triggered by an increase in the concentration of calcium ions in

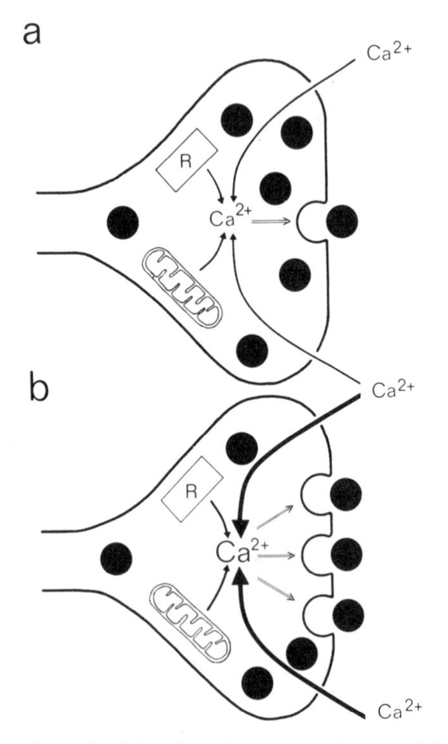

Figure 4.9 Concentration of calcium ions in the cytoplasm and hormonal release by exocytosis. (*a*) Resting condition; small fluctuations in the concentration of calcium ions in the cytoplasm due to the entry of calcium from the extracellular milieu and the release of calcium from mitochondria and internal reservoirs (R) give rise to small amounts of neurohormone release. (*b*) Depolarization of the terminal membrane increases the calcium permeability. The resulting increase in the cytoplasmic concentration of calcium ions results in a large increase in hormone release.

the cytoplasm (figure 4.9). How this causes release is still not clear, but there are several possibilities worth outlining.

Electrostatic effects

As with other vesicles from the cytoplasm, neurosecretory granules carry a net negative charge. For example, granules isolated from the neurohypophysis[31] or adrenal medulla[32] of cattle always migrate to the anode in an electrophoretic field. Since the inner surface of the plasma membrane at the axon ending is almost certainly negatively charged (as in all nerve membranes examined), the granule will be deterred from approaching the surface by electrostatic repulsion. If, now, there is an increase in the concentration of calcium ions (which are strongly positively charged) in the vicinity of the surface membrane, there will be a tendency for the electrostatic repulsion to be much diminished. This might greatly increase the chance of a granule coming into contact with the membrane of the axon ending[33,34] (figure 4.10*a*). The weak point in this argument is that axoplasm is thought to contain magnesium ions at concentrations three or four orders of magnitude higher than that of calcium. It is difficult, therefore, to believe that enough calcium could enter during depolarization to make a significant change in the total concentration of divalent cations at the inner face of the axon membrane. Furthermore, injection of calcium ions into the presynaptic ending of the squid giant axon causes neurotransmitter release, but injection of similar amounts of magnesium ions has no effect.[27] What we are looking for, therefore, is some effect which is more calcium-specific. So, if a reduction in electrostatic repulsion

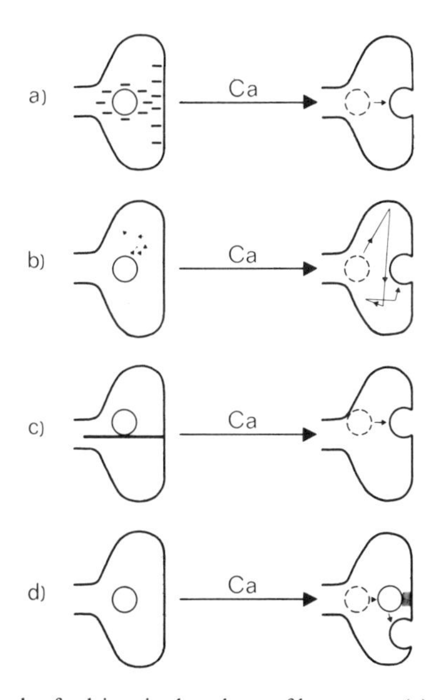

Figure 4.10 Possible role of calcium in the release of hormones. (*a*) Calcium diminishes the electrostatic barrier between the two membranes. (*b*) An increase in the ionized calcium concentration induces a decrease in the cytoplasmic viscosity. (*c*) Calcium ions accelerate the speed at which granules move along microtubules. (*d*) Calcium activates an enzyme which catalyzes the fusion of the granule with the plasma membrane.

is involved, this would mean either that the site of binding producing such a reduction is specific for calcium or a further calcium-specific effect is required after a close approach of the granule to the surface before its contents can be released.

Changes in cytoplasmic viscosity

It has been suggested, that in the case of synapses, entry of calcium ions causes a change in the sol/gel status of the cytoplasm so as to allow the synaptic vesicles to move much more freely by Brownian motion;[35] as a result they collide much more frequently with the surface membrane so that release occurs at a much higher rate (figure 4.10*b*). It has been known for almost thirty years that axoplasm isolated from squid axons can be liquified by calcium ions and, conversely, can be made to gel by treatment with the calcium-chelating compound EGTA. One can easily imagine that a similar chain of events occurs at neurosecretory axon endings. Indeed, there is some evidence that the action potentials of the neurosecretory axons are of unusually long duration (p. 36), and if this were to lead to a correspondingly more prolonged reduction in viscosity, then this might be significant in allowing more time for the relatively large neurosecretory vesicles to move to the cell membrane and fuse with it.

Calcium and microtubules

There is evidence from work on the pancreas that calcium ions may cause the granules containing insulin to move to the surface membrane along microtubules.[36] This might involve the contraction of a microfilament/microtubular complex. It has long been known that microtubules occur in nerve axons, but until recently there has been no evidence that they continue into the endings. However, some recent work suggests that microtubules may, after all, be found in the endings. It is possible, therefore, that the entry of calcium ions might cause neurosecretory granules to move up to the surface membrane much as occurs in insulin release in the pancreas (figure. 4.10*c*). However, the application of alkaloids such as colchicine and vincristine, known to disrupt microtubules, does not affect the release of neurohypophysial hormones evoked by membrane depolarization.[37] This is not to say that movement along microtubules may not be involved in the transport of granules along the axon to maintain the stock of granules in the ending, but it does seem as if microtubules are not involved in the actual process(es) of release.

Possible involvement of nucleotides

Isolated neurosecretory granules ordinarily release their content only after such drastic treatment as osmotic shock in distilled water. Moreover, they are stable in the presence of calcium ions. The inclusion of ATP in the bathing medium does, however, cause the release of the adrenalin from vesicles isolated from the adrenal medulla[38] of the rat and of oxytocin and vasopressin from granules isolated from the rat neurohypophysis.[39] But as this ATP-evoked release occurs even in the absence of calcium ions, it is difficult to know whether the effect is involved in the normal exocytotic process in intact tissue.

So far we have been concerned with the ways in which calcium ions might be involved in the movement of neurosecretory granules up to the surface of the cell. We have seen how calcium ions might cause a reduction of electrostatic repulsion and/or a reduction of cytoplasmic viscosity. However unclear the preliminary steps are, there has finally to be a fusion of the granule membrane with the plasmalemma for exocytosis to occur.

Membrane fusion

Interesting experiments have been made in which two membranes reconstituted from lipids extracted from frogs' brains were brought into contact, and the speed with which they fused was observed. Fusion was much slower in solutions containing only monovalent ions than it was in the presence of divalent ions, and calcium was more effective than magnesium.[40]

The membranes of some neurosecretory granules, such as those isolated from the adrenal medulla, are unusual in that they contain large amounts of the phospholipid, lysolecithin.[41] Other cell inclusions such as mitochondria and microsomes contain only traces of lysolecithin. This compound is very active in causing membrane fusion;[42,43] for example, small amounts of lysolecithin added to a suspension of red blood cells causes them to fuse together to produce extraordinary giant cells. Presumably the presence of lysolecithin in the membranes of neurosecretory granules greatly increases the ease with which they fuse with the surface membrane once they come into contact with it.

We cannot say with any certainty what the exact role of calcium ions is in this process. They may ease membrane fusion as suggested by the experiments with reconstituted membranes, but it is possible that they are only involved in ensuring that the granules reach the surface where their lysolecithin content might be enough to ensure prompt fusion. Unfortunately for this latter view some neurosecretory granules have now been found whose membranes contain rather little lysolecithin.[44] Against this it is claimed that the conversion of the commonly occurring lipid lecithin to lysolecithin is relatively simple, and that such a change might occur as a necessary preliminary or stimulus for fusion. So we are left with no very clear picture as to how membrane fusion is controlled; lysolecithin in the granule membrane and/or calcium ions in the cytoplasm may act to accelerate the process.

There has been some most interesting recent work on exocytosis in rat peritoneal mast cells[45] and the findings are likely also to apply to exocytosis in neurosecretory systems.[46, 52] What has been found strongly suggests that the first event in membrane fusion is a lateral displacement of membrane proteins away from the area of fusion in both plasma and underlying

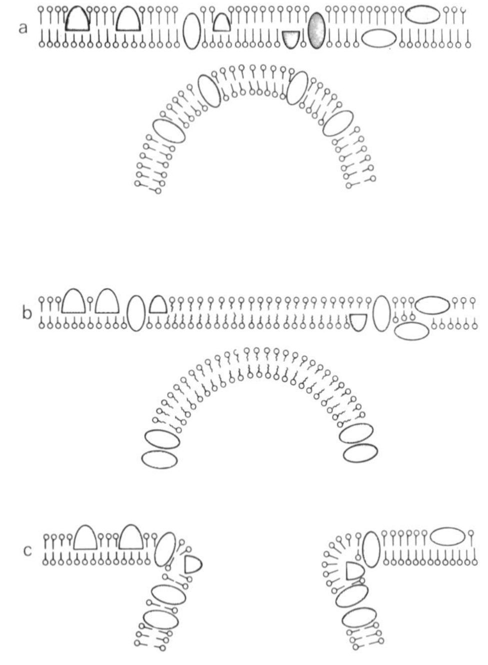

Figure 4.11 Model for membrane fusion. (*a*) The membranes do not fuse because the proteins are randomly arranged in their respective lipid bilayers. (*b*) In membranes in which fusion can occur the lipid molecules have been perturbed. The resulting increase in membrane fluidity gives rise to lateral displacement of the membrane proteins. (*c*) Fusion and fission between the two membranes have occurred (after Ahkong *et al.*, 1975, ref. 50).

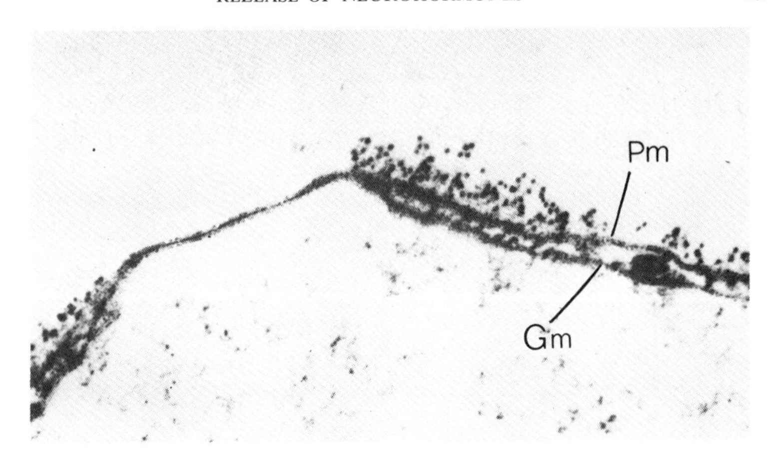

Figure 4.12 Fusion of a secretory granule with the cell membrane in a mast cell stimulated with ovalbumin. The plasma membrane proteins were labelled with a ligand coupled to ferritin, which is electron-dense. Note that the label is bound to all the plasmalemma except where it has fused with the underlying granule membrane. Pm, plasma membrane; Gm, granule membrane (from Lawson *et al.*, 1977, ref. 45).

granule membrane (figures 4.6 and 4.11). As a result of this, displaced proteins are found at higher than normal density at the edges of the fusion sites (figures 4.11 and 4.12). After this displacement, fusion occurs between the two protein-depleted lipid bilayers. It thus seems that for fusion to occur between the two membranes, the proteins have first to be excluded from a small domain in the membranes.

Membrane retrieval

Because exocytosis involves the incorporation into the cell surface of the membranes of the neurosecretory granules, there must be some mechanism for recovering this membrane material or there would otherwise be a great increase in the surface area of the axon ending. The magnitude of the problem will be clear from the following example. In the rat neurohypophysis it has been possible to produce an estimate of the amount of hormone contained in a single granule. From this it can be calculated that, at rest, about 10^9 granules are released per day, while under maximal stimulation, as mentioned earlier, about 10^8 granules are released per minute, i.e. more than a hundred times faster. Since there are about 7×10^7 axon endings in the neural lobe, it follows that, on average, each terminal releases 1 granule every 100 min at rest, rising to 1 granule every 40 s under stimulation. From the surface area of a granule ($8 \times 10^{-2}\ \mu m^2$) and that of a terminal ($7 \mu m^2$), it follows that if there were no membrane retrieval, the plasmalemma of a terminal would, on average, increase in area by about 16% a day even when at rest and, under maximal stimulation, the area of membrane of the terminal would double in an hour!

A widespread feature of neurosecretory axon endings is the presence within them of numbers of small vesicles similar in appearance to synaptic vesicles. The first suggestion that these microvesicles might be involved in the retrieval of excess membrane from the surface came from work on the corpus cardiacum (p. 87) of the blowfly[49] and on the sinus gland of the crab.[47] There was found to be a close relationship between the number of clusters of these vesicles and the occurrence of granules undergoing exocytosis; stimulated preparations contained many more exocytotic profiles and microvesicle clusters than did the resting ones. Significantly, thorotrast particles, widely used as extracellular markers for electron microscopy, have been found to appear in some such microvesicles when the treated neurohaemal tissues have been stimulated. It has therefore been suggested that the microvesicles might be pinched off from the surface membrane by a process similar to pinocytosis (figure 4.13). This mechanism would make sense in that such small vesicles contain the maximum amount of membrane relative to the extracellular fluid taken

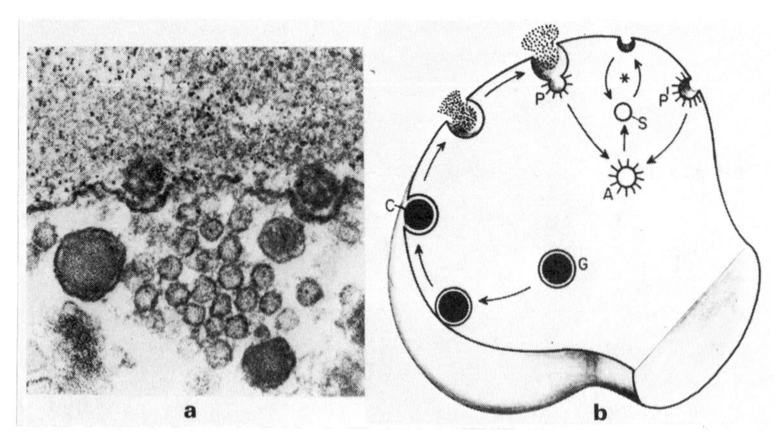

Figure 4.13 (a) Electronmicrograph of neurohormone release in the sinus gland of the crab. (b) Model showing how, after the neurosecretory granule (G) has undergone exocytosis (c), the membrane might be retrieved by microvesiculation. Coated vesicles are formed at the site of release (P) or at other sites along the surface membrane (P¹). Coated vesicles (A) within the endings might lose their coat and give rise to microvesicles (S) which can also be formed directly from the cell membrane (from Bunt, 1969, ref. 47).

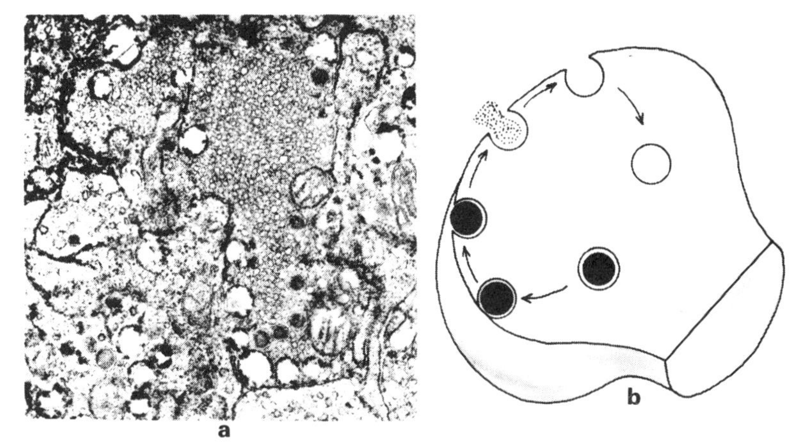

Figure 4.14 Model showing how, after exocytosis, the membrane might be retrieved as vacuoles of the same size as that of the neurosecretory granules (*a*, from Nordmann *et al.*, 1974, ref. 48; *b*, from Nordmann, 1979, ref. 5).

up. It would also have the advantage that the volume of the small vesicles involved would be less than that of the single large granule whose contents were released, and this would allow replacement neurosecretory granules to enter the terminal.[8]

As attractive as this hypothesis is, recent experiments on stimulated rat neurohypophyses both *in vivo* and *in vitro* and on the crab sinus gland *in vitro* have shown that much more horseradish peroxidase is, in fact, taken up from the extracellular space into considerably larger electron-lucent vacuoles of a size very similar to that of the neurosecretory granules themselves[5,48] (figure 4.14). In addition, in rats it has been possible to label the membranes of the neurosecretory granules of the neurohypophysis by injecting radioactive choline into the brain (3rd ventricle). Because newly synthesized granules pass down the axons very much faster than does the flow of axonal membrane, the only membranes in the axon endings which become labelled to any significant extent are those of the neurosecretory granules. After stimulating release of vasopressin by feeding the rats with 2% NaCl solution, and subfractionation of the neural lobe, the label was not found to be associated with the pool of microvesicles but instead remained in the fraction containing the neurosecretory granules.[3] This latter fraction very probably contains the above-mentioned electron-lucent vacuoles, although so far the experiments necessary to show whether their membranes become labelled have not been done. In other experiments, careful examination of electronmicrographs of stimulated and unstimulated rat neurohypophysis and crab sinus glands using the technique of stereology*, which has only recently been applied to biological problems, has shown that after hormone release there is in the axon endings a decrease in the number of neurosecretory granules and an increase in the number of electron-lucent vacuoles. The population of microvesicles did not change significantly[49]. Finally, experiments with radioactive extracellular markers such as mannitol, albumin and inulin show that these molecules are taken up into the axon endings when hormone release is stimulated.[48] From the amounts of these markers found in the tissue it was clear that the volume of extracellular fluid taken up was close to that of the neurosecretory granule content released. It is worth adding that this uptake was blocked when calcium entry was prevented during depolarization of the endings. This suggests that membrane retrieval is a consequence of secretion, and cannot be stimulated by depolarization alone.

Clearly there is a good deal of evidence to support the proposal that membrane retrieval is achieved in many cases by uptake in the form of relatively-large electron-lucent vacuoles. On the other hand, as we saw earlier, in some tissues such as the corpora cardiaca of the blowfly, it is thought that the microvesicles are a significant vehicle for membrane retrieval. One possible explanation of this divergence is that two routes of membrane retrieval exist whose relative importance may depend on the tissue examined and possibly also on the intensity of secretory activity. Alternatively, it is not impossible that one of the apparently different mechanisms is actually an earlier stage in the other; small vesicles might fuse to give larger ones, or larger vacuoles might break down to form a number of smaller vesicles.

REFERENCES

1. Morris, J. F. and Cannata, M. A. (1973) "Ultrastructural preservation of the dense core of posterior pituitary neurosecretory granules and its implication for hormone release," *J. Endocr.* **57,** 517–529.
2. Trifaro, J. M., Poisner, A. M. and Douglas, W. W. (1967) "The fate of the chromaffin granules during catecholamine release from the adrenal medulla. I. Unchanged efflux of phospholipids and cholesterol," *Biochem. Pharmac.* **16,** 2095–2100.
3. Swann, R. W. and Pickering, B. T. (1976) "Incorporation of radioactive precursors into the membrane and contents of the neurosecretory granules of the rat neurohypophysis as a method of studying their fate." *J. Endocr.* **68,** 95–108.
4. Normann, T. C. (1976) "Neurosecretion by exocytosis," *Int. Rev. Cytol.* **46,** 1–77.
5. Nordmann, J. J. (1979) "Hormone release and membrane retrieval in neurosecretion," pp. 619–636, in *Cell Biology of Hypothalamic Neurosecretion*, Eds. J. D. Vincent and C. Kordon, Paris, CNRS.

* Stereology, as used in ultrastructural investigations, involves techniques designed to enable one (*a*), from randomly selected micrographs, to make deductions as to the true three-dimensional form of structures, part only of which may appear in single sections, and (*b*) to make estimates of the numerical density of populations of specific elements present in the structure examined.

6. Diner, O. (1967) "L'expulsion des granules de la medullo-surrenale chez le hamster," *C. r. hebd. Séanc. Acad. Sci.*, Paris, **265,** 616–619.
7. Weiss, M. (1965) "The release of pituitary secretion in the platyfish, *Xiphophorus maculatus* (Guenther)," *Z. Zellforsch mikrosk. Anat.* **68,** 783–794.
8. Douglas, W. W. (1974) "Mechanism of release of neurohypophysial hormones: stimulus-secretion coupling," *Handbook Physiol.*, Vol. IV, pp. 191–224, Am. Physiol. Soc., Washington.
9. Normann, T. C. (1969) "Experimentally induced exocytosis of neurosecretory granules," *Expl. Cell. Res.* **55,** 285–7.
10. Brady, J. and Maddrell, S. H. P. (1967) "Neurohaemal organs in the medial nervous system of insects," *Z. Zellforsch. mikrosk. Anat.* **76,** 389–404.
11. Maddrell, S. H. P. (1966) "The site of release of the diuretic hormone in *Rhodnius*—a new neurohaemal system in insects," *J. exp. Biol.* **45,** 499–508.
12. Bretscher, M. S. & Whytock, S. (1977) "Membrane-associated vesicles in fibroblasts," *J. Ultrastruct. Res.* **61,** 215–217.
13. Smith, A. D. , De Potter, W. P., Moerman, E. J., and Schaepdryver A. F. (1970) "Release of dopamine β-hydroxylase and chromogranin A upon stimulation of the splenic nerve," *Tissue and Cell.* **2,** 547–568.
14. Cheng, K. W., Martin, J. B. and Friesen, H. G. (1972) "Studies on neurophysin release," *Endocrinology*, **91,** 177–184
15. Fawcett, C. P., Powell, A. E. and Sachs, H. (1968) "Biosynthesis and release of neurophysin," *Endocrinology* **83,** 1299–1310.
16. Nordmann, J. J., Dreifuss, J. J. and Legros, J. J. (1971) "A correlation of release of 'polypeptide hormones' and of immunoreactive neurophysin from isolated rat neurohypophyses," *Experientia* **27,** 1344–1345.
17. Uttenthal, L. O., Livett, B. G. and Hope, D. B. (1971) "Release of neurophysin together with vasopressin by a Ca^{2+}-dependent mechanism," *Phil. Trans. Roy. Soc. Lond. B.* **261,** 379–380.
18. Edwards, B. A., Edwards, M. E. and Thorn, N. A. (1973) "The release *in vitro* of vasopressin unaccompanied by the axoplasmic enzymes: lactic acid dehydrogenase and adenylate kinase," *Acta. Endocrinol.* **72,** 417–424.
19. Matthews, E. K., Legros, J. J., Grau, J. D., Nordmann, J. J. and Dreifuss, J. J. (1973) "Release of neurohypophysial hormones by exocytosis," *Nature* **241,** 86–88.
20. Schober, R., Nitsch, C., Rinne, U., and Morris S. J. (1977) "Calcium-induced displacement of membrane-associated particles upon aggregation of chromaffin granules," *Science* **195,** 495–7.
21. Douglas, W. W. and Rubin, R. P. (1961) "The role of calcium in the secretory response of the adrenal medulla to acetylcholine," *J. Physiol.* **159,** 40–57.
22. Douglas, W. W., and Poisner, A. M. (1964) "Stimulus-secretion coupling in a neurosecretory organ: the role of calcium in the release of vasopressin from the neurohypophysis," *J. Physiol.* **172.** 1–18.
23. Berlind, A. and Cooke, I. M. (1968) "Effect of calcium omission on neurosecretion and electrical activity of crab pericardial organs," *Gen. Comp. Endocr.* **11,** 458–463.
24. Maddrell, S. H. P., and Gee, J. D. (1974) "Potassium-induced release of the diuretic hormones of *Rhodnius prolixus* and *Glossina austeni*: Ca dependence, time course and localization of neruohaemal areas," *J. exp. Biol.* **61,** 155–171.
25. Nordmann, J. J. and Currell, G. A. (1975) "The mechanism of calcium ionophore-induced secretion from the rat neurohypophysis," *Nature* **253,** 646–647.
26. Nakazato, Y., and Douglas, W. W. (1974) "Vasopressin release from the isolated neurohypophysis induced by a calcium ionophore, X-537A," *Nature* **249,** 479–481.
27. Miledi, R. (1973) "Transmitter release induced by injection of calcium into nerve terminals," *Proc. R. Soc. B.* **183,** 421–425.
28. Douglas, W. W. and Poisner, A. M. (1964) "Calcium movements in the neurohypophysis of the rat and its relation to the release of vasopressin," *J. Physiol.* **172,** 19–30.

29. Baker, P. F., Meves, H. and Ridgway, E. B. (1973) "Effects of manganese and other agents on the calcium uptake that follows depolarization of squid axons," *J. Physiol.* **231,** 511–526.
30. Dreifuss, J. J., Grau, J. D. and Nordmann, J. J. (1973) "Effects on the isolated neurohypophysis of agents which affect the membrane permeability to calcium," *J. Physiol.* **231**, 96P–98P.
31. Vilhardt, H. and Jorgensen, O. (1972) "Free flow electrophoresis of isolated secretory granules from bovine neurohypophyses," *Experientia* **28,** 852.
32. Banks, P. (1966) "An interaction between chromaffin granules and calcium ions," *Biochem. J.* **101,** 18c–20c.
33. Dean, P. M. (1975) "Exocytosis modelling: an electrostatic function for calcium in stimulus-secretion coupling," *J. theor. Biol.* **54,** 289–308.
34. Dean, P. M. and Matthews, E. K. (1975) "The London-Van der Waals attraction constant of secretory granules and its significance," *J. theor. Biol.* **54,** 309–321.
35. Shaw, T. I. and Newby, B. J. (1972) "Movement in a ganglion," *Biochim. biophys Acta.* **255,** 411–422.
36. Malaisse, W. J., Malaisse-Lagae, F., Walker, M. O. and Lacy, P. E. (1971) "The stimulus-secretion coupling of glucose-induced insulin release. V. The participation of a microtubular-microfilamentous system," *Diabetes* **20,** 257–265.
37. Rufener, C., Orci, L., Nordmann, J. J. and Rouiller, Ch. (1972) "Effect of vincristine on rat neurohypophysis *in vitro*," *Gen. Comp. Endocrin.* **18,** 621.
38. Poisner, A. M. and Trifaro, J. M. (1967) "The role of ATP and ATP-ase in the release of catecholamines from the adrenal medulla. I. ATP-evoked release of catecholamines, ATP and protein from isolated chromaffin granules," *Mol. Pharmacol.* **3,** 561–571.
39. Poisner, A. M. and Douglas, W. W. (1968) "A possible mechanism of release of posterior pituitary hormones involving adenosine triphosphate and an adenosine triphosphatase in the neurosecretory granules," *Mol. Pharmacol.* **4,** 531–540.
40. Blioch, Z. L., Glagoleva, I. M., Liberman, E. A. and Nenashev, V. A. (1968) "A study of the mechanism of quantal transmitter release at a chemical synapse," *J. Physiol.* **199.** 11–35.
41. Winkler, H. (1971) "The membrane of the chromaffin granule," *Phil. Trans. Roy. Soc. Lond. Ser. B.* **261,** 293–303.
42. Poole, A. R., Howell, J. I. and Lucy, J. A. (1970) "Lysolecithin and cell fusion," *Nature* **227,** 810–813.
43. Lucy, J. A. (1970) "The fusion of biological membranes," *Nature* **227,** 814–817.
44. Vilhardt, H. and Hølmer, G. (1972) "Lipid composition of membranes of secretory granules and plasma membranes from bovine neurohypophyses," *Acta. Endocrinol.* **71,** 638–648.
45. Lawson, D., Raff, M. C., Gomperts, B., Fewtrell, C. and Gilula, N. B. (1977) "Molecular events during membrane fusion. A study of exocytosis in rat peritoneal mast cells," *J. Cell. Biol.* **72,** 242–259.
46. Theodosis, D., Dreifuss, J. J. and Orci, L. (1978) "A freeze-fracture study of membrane events during neurohypophysial secretion", *J. Cell Biol.* **78,** 542–553.
47. Bunt, A. H. (1969) "Formation of coated and 'synaptic' vesicles within neurosecretory axon terminals of the crustacean sinus gland," *J. Ultrastruct. Res.* **28,** 411–421.
48. Nordmann, J. J., Dreifuss, J. J., Baker, P. F., Ravazzola, M., Malaisse-Lagae, F. and Orci, L. (1974) "Secretion-dependent uptake of extracellular fluid by the rat neurohypophysis," *Nature* **250,** 155–157.
49. Nordmann, J. J. and Morris, J. F. (1976) "Membrane retrieval at neurosecretory axon endings," *Nature* **261,** 723–725.
50. Ahkong, Q. F., Fisher, D., Tampion, W. and Lucy, J. A. (1975) "Mechanism of cell fusion," *Nature* **253,** 194–195.
51. Smith, A. D. (1971) "Secretion of proteins (chromogranin A and dopamine-β-hydroxylase) from a sympathetic neuron," *Phil. Trans. Roy. Soc. Lond. B,* **261,** 363–370.
52. Aunis, D., Hesketh, J. E. and Devilliers, G. (1979) "Freeze-fracture study of exocytosis in bovine adrenal medulla", *Cell Tiss. Res.*, in press.

CHAPTER FIVE

SITES OF NEUROHORMONE RELEASE

Neurohaemal areas and organs; the sites of neurohormone release

We have so far been concerned largely with those features of neurosecretory cells that are common to all the animals that possess them—they all synthesize an active principle in their cell bodies, transport it to the release area(s), and liberate it into the extracellular fluid. Animals differ very much, however, in the distribution and number of their neurosecretory cells and in how the cells are arranged so as to achieve their requisite effects. For example, in many invertebrate groups there is little or no circulation of the extracellular fluid, and this must, of course, affect the ways in which neurosecretory cells can be used to control other different organs, epithelia or cells. In some animals able to live in a variety of environments, the blood composition and concentration varies within quite wide limits. The complex central nervous systems of more advanced animals seem to require working conditions controlled more finely than this. It is not surprising, therefore, that in evolution these animals have evolved with blood-brain barriers able not only to restrict movement between the general extracellular environment and the milieu in which most of the nerve cells operate, but also to regulate more closely the composition of the fluid bathing the cells in the central nervous system.[1] To take a simple example, nerve cells in the ventral nerve cord of the cockroach are just as conservative and conformist as those in most animals in that they require a sodium-rich potassium-poor medium in order to transmit nerve impulses. So effective is the blood-brain barrier, however, that an isolated cockroach nerve cord continues to function for some hours, even when bathed in a fluid completely lacking in sodium ions.[2] Should the surface of a nerve cord in such a solution be damaged, however, the nerve cells lose their ability to conduct within about a minute. The possession of such a blood-brain barrier system has obvious advantages, but it is clearly something of an embarrassment to neurosecretory cells which lie behind it. How are they to liberate their active materials into the general extracellular fluid? The simple solution would be for branches of their axons to run out through the blood-brain barrier. However, each such crossing would introduce a potential leak in the barrier and, as we have seen (p. 24), each neurosecretory cell may have as many as four thousand axon terminals. How this problem is overcome is well illustrated by the adaptations of neurohaemal systems of insects, crustaceans and vertebrates described on pp. 87, 92 and 97 respectively.

In this chapter we shall illustrate two main themes. First, because of the somewhat patchy state of knowledge of neurosecretory systems in many animals, we shall survey the relevant information on these systems in different invertebrate and vertebrate groups in an attempt to show the extent to which the different classes make use of neurosecretion. Second, we shall attempt to interpret the morphology and mode of operation of each system in terms of the function it has to fulfil and of the limitations put on it by the functioning of other systems—such as the extent and speed of circulation of the extracellular fluid, and the existence or not of a blood-brain barrier. In other words, we shall look to see how the system is designed to maximize its effectiveness.

Coelenterates

Among the nerve cells found in *Hydra* are some for which there is good evidence that they are neurosecretory. They contain secretory droplets which can be stained with vital dyes and with paraldehyde fuchsin (PAF) which is known to have a high affinity for neurosecretory material.[3] Ultrastructurally, they are found to contain granules similar to elementary neurosecretory granules and which arise from the Golgi apparatus. Finally, a substance which activates head and bud formation can be isolated from the nerve cells,[4] and from that fraction of homogenates of the animal which contains granules similar in appearance to those seen in the neurosecretory cells.

The neurosecretory cells are most numerous towards the anterior end of the animal just proximal to the hypostome (figure 5.1). The cell processes are relatively short and end in contact with the mesogloea, the extracellular material lying between the two epithelial layers of the animal, i.e. the main extracellular space. Substances released by these cells will thus be in the right position to diffuse in the extracellular space, provided of course that the endodermal and ectodermal epithelia are not so permeable that the substances are swiftly lost through them into the gut or to the outside. Since if a *Hydra* is ligatured below the hypostome it steadily swells when placed in fluids of low osmotic strength, it is clear anyway that the body wall

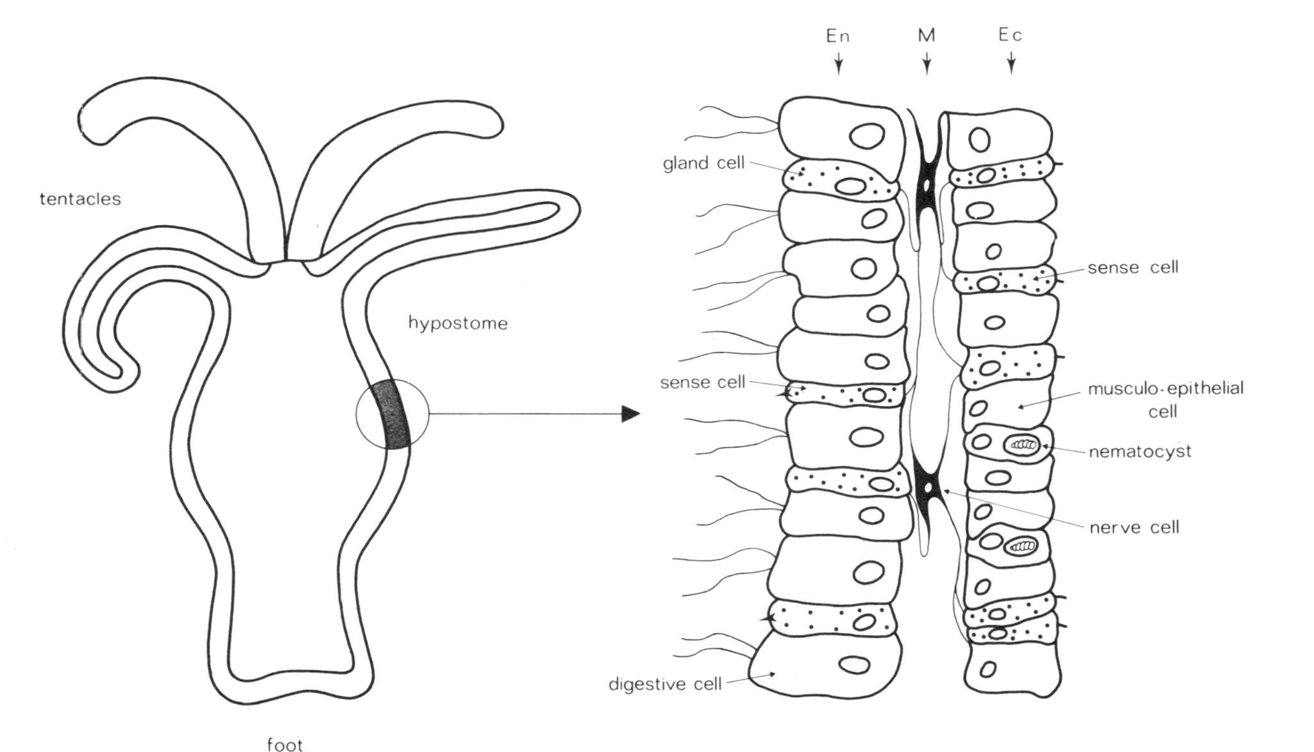

Figure 5.1 Representations of longitudinal sections through a hydra to show the position of the nerve cells, a proportion of which are neurosecretory, M—mesogloea; En—endoderm; Ec-ectoderm.

(consisting of the endodermal and ectodermal layers) is sufficiently impermeable to substances in the enteron to allow them to exert an osmotic effect across the wall. Unless the two layers have very different permeabilities, it is likely, therefore, that the mesogloea can be looked upon as a true internal compartment, and the position of the neurosecretory cell endings in the mesogloea makes sense. Because the mesogloea is not a fluid-filled space but contains fibrous elements set in a viscous medium, substances in it must rely nearly entirely on diffusion for their circulation. At first sight this would appear to be something of a difficulty, but it is a feature which in fact appears to have been turned to advantage. It seems that it may well be the case that the diffusion of neurosecretory substance(s) released predominantly at the head end sets up a concentration gradient along the length of the animal, and that this gradient is used to control the morphogenetic fate of the cells lying in the different positions in the gradient. This would clearly not be possible if the extracellular fluid were more mobile and were more rapidily circulated. Since *Hydra* is a small animal, the ratio between the surface area of its neurosecretory cells and the volume of the extracellular fluid is automatically greater than in larger animals. It is therefore not surprising that one does not find the large numbers of neurosecretory cells, each with many axon endings, that are found in bigger animals or in animals with extensive volumes of extracellular fluid. However, the purified head-activating substance is a very active compound,[4] being effective at concentrations of less than 10^{-10}M. In *Hydra*, then, it seems to be the case that the problem of how to affect other cells by liberating an active compound into extracellular fluid is overcome by producing a highly active substance rather than by modifying the releasing cells to produce large amounts of material.

Platyhelminthes

Although there are several studies implicating the neurosecretory cells of platyhelminthes in the control of a variety of functions, there is little one can say about the structural organization of the neurosecretory system. Neurosecretory cells are known in all three major groups of the phylum. In turbellarians and trematodes, there are neurosecretory cells in the hind part of the brain, and these cells stain intensely with PAF and contain dense membrane-bound granules.[5] In most cases it is not clear, however, where the axon endings are. In schistosomes, the axons run close to muscle and other cells, and the axons contain, in addition to dense neurosecretory granules, what are referred to as *clear axoplasmic vesicles*.[6] If these are not due to fixation at too alkaline a pH (which as we have seen on p. 57 causes the granules to lose their electron density), then they might be the products of membrane retrieval after release, i.e. they may be the electron-lucent vacuoles which recent research suggests are the main way in which granule membrane is recovered.

As with *Hydra*, the extracellular fluid in platyhelminths is not extensive, so that one would not expect the elaboration of the neurosecretory system that one finds elsewhere. Also neurosecretion in platyhelminths is (so far) implicated only in controlling relatively slow changes in growth and development, so the system may not need to be able to release neurosecretory substances rapidly; its lesser development may be sufficient for the needs of these animals.

Nematoda

Neurosecretory cells, staining with PAF, have been found in *Ascaris lumbricoides* and in *Phocanema decipiens*.[7] The cells do not stain as intensely as do, say, those of insects, so it has been suggested that transport away from the cell bodies might be faster relative to the rate of granule formation. In these two nematodes the axons were not easily visible; and no obvious centre for the release of the neurosecretory material could be seen. In *Haemonchus contortus*, axons containing numerous neurosecretory granules were found close behind the excretory gland and the excretory pore. They appeared to contain "empty" neurosecretory granules, but the axons also appeared to merge without limiting membranes into a so-called light area beneath the cuticle, so that fixation may not have been optimal. However, there is some evidence of neurosecretory control of the excretory gland, so that it is possible that the empty neurosecretory granules were the products of membrane retrieval as electron-lucent vacuoles.

Neurosecretion in animals lacking circulatory systems

Before we go on to consider the neurosecretory systems of animals having a more complex "higher" level of organization, it is worth pausing to consider whether there are features special to the neurosecretory systems of coelenterates, platyhelminths and nematodes. The animals of these groups have little in the way of circulatory systems, yet some of them are quite large in size. They cannot, therefore, rapidly raise the concentration of a substance in the whole extracellular fluid space. Thus it may be significant that neurosecretion in these animals appears to be used to control only the longer-term changes involved in morphogenesis and reproduction rather than the rapid metabolic changes induced by the neurohomones of animals with a well-developed ability to circulate the extracellular fluid.

Polychaete annelids

Most of what is known about the organization of the neurosecretory system in polychaete annelids is based on studies on nereids, in particular their cerebral ganglia. Although neurosecretory cell bodies have been found in the ventral nerve cord of several polychaete species, nothing is known of how, where, or even whether, they release material into the blood or other extracellular fluids. The cerebral ganglia contain quite large numbers of neurosecretory cells of a range of different types (based on differential staining reactions). The axons of these cells aggregate in tracts, and the tracts run ventrally to the brain floor where they end in contact with the neural lamella.[8] The endings are of two types, one characteristic of neurosecretory axon endings with many close-packed elementary neurosecretory granules and another, which had become known as secretory end-feet, where the endings are larger, with fewer granules, but with large numbers of distinctive mitochondria.

Closely applied to the ventral surface of the brain, where the neurosecretory axons end, is an epithelial structure of probable endocrine function, known as the *infracerebral gland* (figure 5.2). Among its epithelial cells are cells similar in appearance to neurosecretory cells; they contain material that stains with such neurosecretory stains as paraldehyde fuchsin, and they have cellular processes like short axons. Although there is no evidence that the secretory cells of the infracerebral glands are nervous in character, they are reminiscent of the neurosecretory cell bodies one finds in corpora cardiaca, the cerebral neurohaemal organs of insects, which cells also lie outside the central nervous system. Some of the neurosecretory axons from the brain cross the neural lamella separating the brain from the infracerebral gland and form terminals within the gland; again this has analogies with the corpora cardiaca of insects. The situation is summarized in figure 5.2.

The presence close to the supposed neurohaemal areas of both a part of the coelomic cavity and of a blood vessel would seem to allow hormones (which might be released) access to either the coelomic fluid, or the blood, or both. Hormonal effects can be produced by implantation of cerebral ganglia into the coelom, so that the coelomic fluid can evidently act to distribute hormonally active materials. It is not known whether hormones

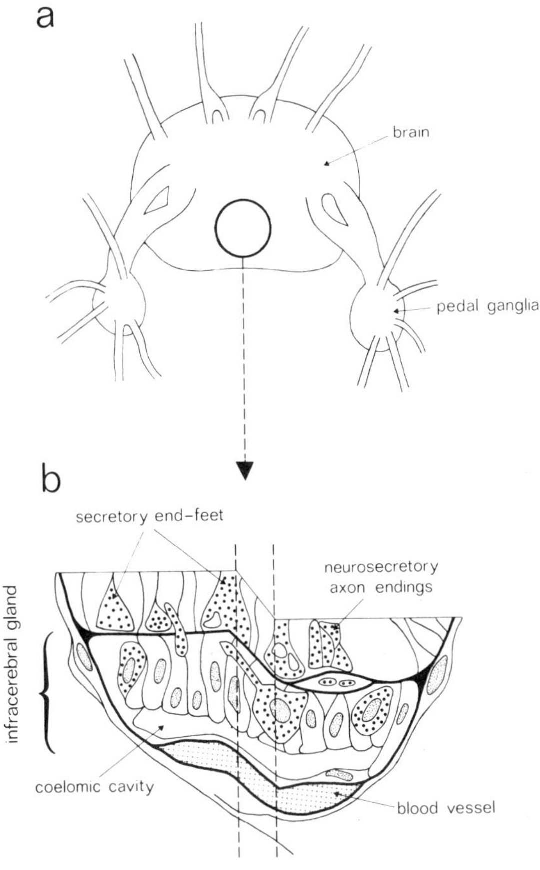

Figure 5.2 To show (*a*) the position and (*b*) the structure of the infracerebral gland in a nereid polychaete ((*b*) after Baskin (1976), ref. 8)

are circulated in the blood stream, but in contrast to the central nervous system of oligochaetes, that of polychaetes is only poorly vascularized, so perhaps the coelomic fluid is the more important vehicle for the transport of released hormones.

Other annelids

As with polychaetes, the cerebral ganglia of oligochaetes and leeches contain many neurosecretory cell bodies of different types, separable by their staining properties, or by the size and appearance of the elementary

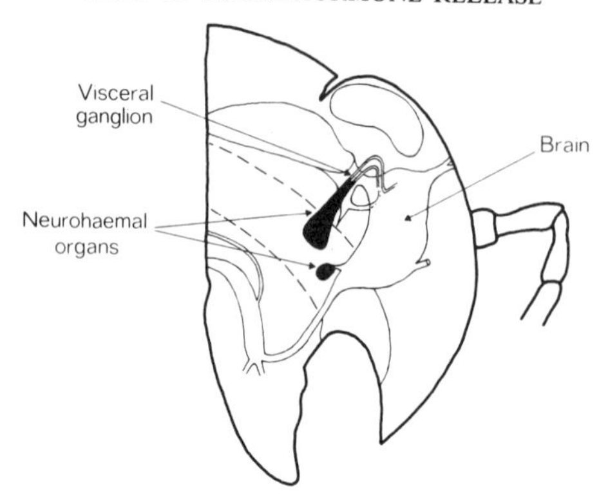

Figure 5.3 The neurohaemal organs of the head in a myriapod. The upper structures are the lateral cerebral organs and the lower ones, the lateral connective bodies (after Prabhu (1961). ref.9).

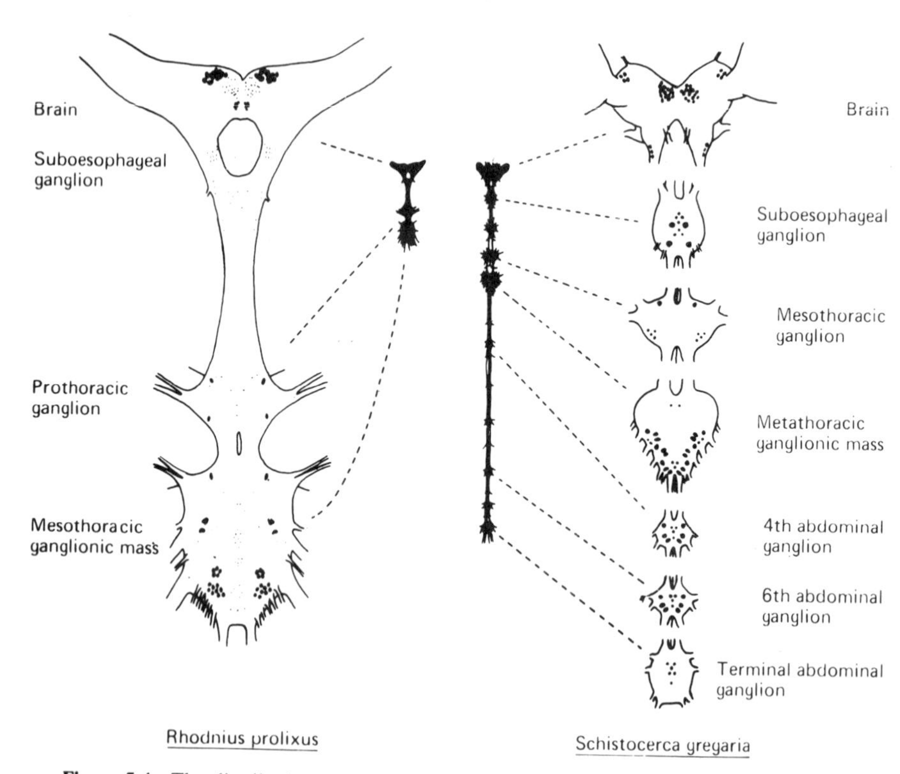

Figure 5.4 The distribution of neurosecretory cells (dots) in the central nervous system in two insects (from Maddrell (1974), ref. 10).

neurosecretory granules that they contain. The central nervous system of oligochaetes is well vascularized, and there is some suggestion that the neurosecretory axons may terminate on the blood capillaries, but this has yet to be definitely established. In leeches, the neurosecretory axons terminate at the posterior surface of the brain in the mid line. In both leeches and oligochaetes, neurosecretory cell bodies occur in the ventral nerve cord, but details of their morphology are lacking.

Myriapods

In diplopodan myriapods, neurosecretory cell bodies have been found at various sites in the brain.[9] Close to each optic lobe there is a lateral cerebral organ (figure 5.3) which contains apparently glandular cells as well as neurosecretory axons thought to originate in the frontal lobes of the brain. In some myriapods there are, in addition, lateral connective bodies (figure 5.3) which also appear to be neurohaemal structures.[9] These cerebral organs seem to have similarities with the corpora cardiaca-corpora allata complexes of insects (p. 90). This is not surprising as the myriapods are thought to be closely related to the insects.

Insects

The distribution of neurosecretory cells in the insect central nervous system has been studied in a wide variety of species (see ref. 10 for example). In all of them there are found to be large numbers of neurosecretory cells of different types (as judged by their differing staining reactions) in all the ganglia of the nerve cord. Figure 5.4 shows the distribution found in two typical cases.

Insects maintain the cells of the central nervous system in an environment which is more closely regulated than is the haemolymph (blood). This control is exerted by a sheath of specialized cells called the *perineurium* which envelops the whole of the central and much of the peripheral nervous system.[2] We shall see below how the neurosecretory cells which are thereby apparently separated from the haemolymph overcome the problem of releasing their active products into it.

In addition to neurosecretory cells in sites in the central nervous system there are also reports of neurosecretory cell bodies lying on the peripheral nerves[11] (figure 5.5). These cells differ in that they are multipolar with axons running in more than one direction. As with other neurosecretory

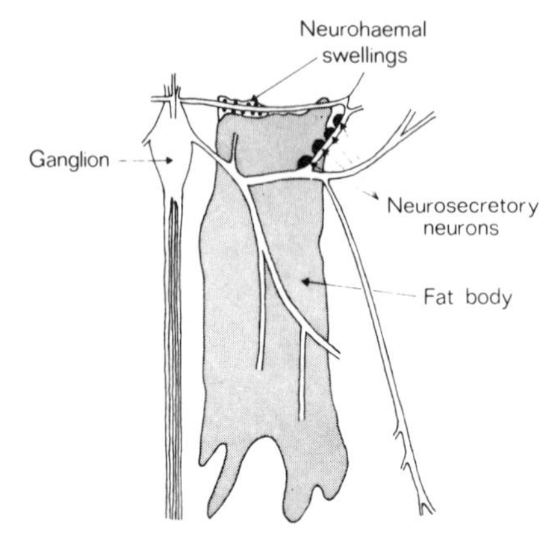

Figure 5.5 Peripheral neurosecretory cells in the abdomen of the stick insect *Carausius morosus*.

cells, the cell bodies contain electron-dense granules. Unusually, however, they and most of the length of their axons are not maintained in an environment distinct from the haemolymph, but are freely exposed to the haemolymph. In the stick insect *Carausius morosus*, this would seem to raise a difficulty as the haemolymph contains less than 10 mM sodium. Under the circumstances it is perhaps not surprising that the inward current in the action potentials of both cell bodies and axons is largely carried by calcium ions.[12,13]

In most insects, the neurosecretory cells release their active products into the haemolymph, and for those protected by the perineurium there is thus a problem. For access to the haemolymph the neurosecretory axons must cross the perineurium. (An alternative solution would be to release their products inside the central nervous system and rely on diffusion or transport through the perineurium; no cases are known where this occurs.) To cross the perineurium and yet not disturb the controlled environment of the nerve cells, the solution most often arrived at is as follows. The axons from a number of neurosecretory cells collect together as a tract and run away from the central nervous system for a short distance in one of the branches of the nerves of the sympathetic or visceral system leaving the particular ganglion. Only then do they all emerge together from underneath the perineurium. In this way the contact between the haemolymph and the extracellular environment of the nerves, particularly their cell bodies, is minimized.

For a neurosecretory cell to release sufficient active material into as large a volume as the haemolymph, its axon must branch profusely and this is found to be the case. Figure 5.6 shows the structure of a neurohaemal organ of the wasp *Vespula,* with a small number of neurosecretory axons leaving the nerve and branching so often that the large mass of axon endings forms an easily discernible swelling on the nerve.[14]

Once free of their perineurial ensheathment, the axon endings are in close contact with the haemolymph. In this way the materials they release have free access to the general circulation.

Further modifications found in insect neurohaemal organs are that the acellular nerve sheath runs deeply into the organ at many places as a stroma, and that often there are, in addition, deep invaginations of the haemocoel (blood space) into the organ. Both these arrangements allow the area of contact between the axon endings and the haemolymph to be greatly increased with very little increase in the space occupied by the organ. Many neurohaemal organs contain not only axon endings but also intrinsic neurones which are often themselves neurosecretory in type.

It is worth noting that in those insects where there is a continuous fat

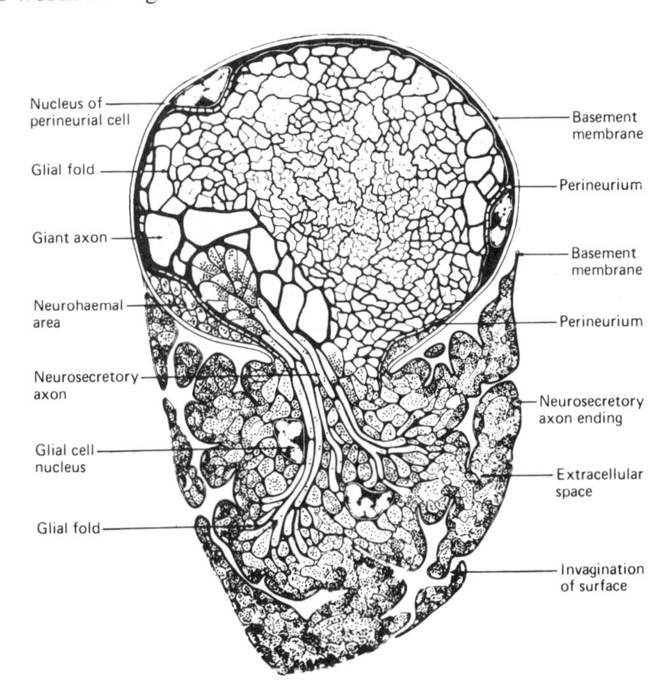

Figure 5.6 Transverse section through a neurohaemal organ on a transverse nerve of the sympathetic nervous system of the wasp *Vespula germanica* (after Raabe *et al.*, 1970, ref. 14, from Maddrell, 1974, ref. 10).

body layer round the central nervous system, it is missing in the immediate vicinity of the neurohaemal organs.[15] Obviously this allows substances released from the organs more direct access to the general body of circulating haemolymph.

As is clear from figure 5.4, there are significantly more neurosecretory cells in the insect brain than in any of the other ganglia. This may explain why the *corpora cardiaca*, the neurohaemal organs of the neurosecretory cells of the brain, have been known for almost fifty years, while the neurohaemal organs of the other ganglia have only been recognized during the last ten years.[16] Figure 5.7 shows the segmental organization of neurohaemal organs in a typical insect. In certain orders, notably Coleoptera, neurohaemal organs are missing in the thoracic region, but in others they are well developed all along the entire ventral nerve cord. The functional significance, if any, of these anatomical differences is not clear.

The fact referred to above (that neurohaemal organs may contain intrinsic neurosecretory cells as well as axons from nearby ganglia) makes it difficult to decide which cells are responsible for producing a particular

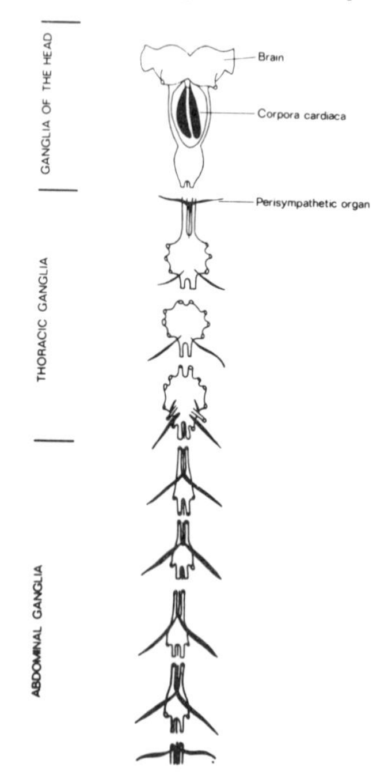

Figure 5.7 The neurohaemal organs (shown in black) of a typical insect (after Raabe *et al.*, 1971, ref. 16).

hormone occurring in a neurohaemal organ. An interesting technique can be used to unravel the situation. The nerves between the corpora cardiaca and the brain are cut and the corpora cardiaca removed. The cut ends of the nerves then heal over and regenerate new neurohaemal organs, termed *de novo* corpora cardiaca. These bodies lack the intrinsic neurosecretory cells of normal corpora cardiaca, and so assay of them is not affected by the contents of such intrinsic cells. Equally, the only hormones released by, say, feeding are those produced by cells whose perikarya (cell bodies) lie in the brain. This technique has been used, for example, to show that the hyperglycaemic and diuretic hormones of the locust *Locusta migratoria* are secreted by neurosecretory cells in the brain, but that the adipokinetic hormone is secreted by cells intrinsic to the corpora cardiaca.[17]

Although most neurosecretory axons run to, and end in, neurohaemal areas or organs, many cases are known where the nerves supplying various organs are found to contain axons which ultrastructurally appear to be neurosecretory. It is tempting to suggest, of course, that these axons act to control the particular organs towards which the axons run. One must be very cautious, however, before accepting such an interpretation. It turns out that some neurosecretory axons which release their products into the haemolymph do so after running considerable distances along peripheral nerves. Only if a neurosecretory axon penetrates beneath the basement membrane of a particular organ, loses its glial and/or perineurial wrapping and comes into close contact with the cells of the organ, can it be said that there is good morphological evidence of neurosecretory control of the organ. Some neurosecretory axons are found to satisfy these criteria, so that it seems very likely that they do have a role in the control of the organ. Because of the localized nature of the release, it is arguable whether such control can be termed hormonal. The term *neurosecretomotor control* has been suggested as a more appropriate description. It is in areas like this that we see the difficulty of classifying an axon as neurosecretory rather than as one releasing a neurotransmitter. This difficulty really arises because all nerves exert their effects on other cells in basically the same way—by the release of active chemical substances.

We have seen that insect neurosecretory cells have profusely branched axons, presumably to increase the amount of hormone that can be released. This is not the only way in which this might be achieved, however, and one finds also that the number and size of neurosecretory cell bodies is greater in larger insects. It has been calculated,[18] for instance, that the volume of the medial neurosecretory cells in the tiny fruit fly *Drosophila,* and in large Saturnid moths differs by a factor of 300–400. A ranking of a range of species in order of the calculated volume of their neurosecretory cells fits well with a list of the same species arranged in order of increasing size. This is in marked contrast to the situation with other neurones. For example, the locust, the bumble-bee and the blowfly all have about the same number of flight motor neurones, each of about the same size.

Crustacea

As in insects, the brain and all the ganglia of the ventral nerve cord contain neurosecretory cells. In addition, there are neurosecretory cell bodies in the optic ganglia which lie anterolateral to the brain; these groupings of cells are called the X-organs (see figure 5.8). The environment of most marine crustaceans is, of course, very stable in composition and, since their blood is basically similar to sea water, it would seem that these animals

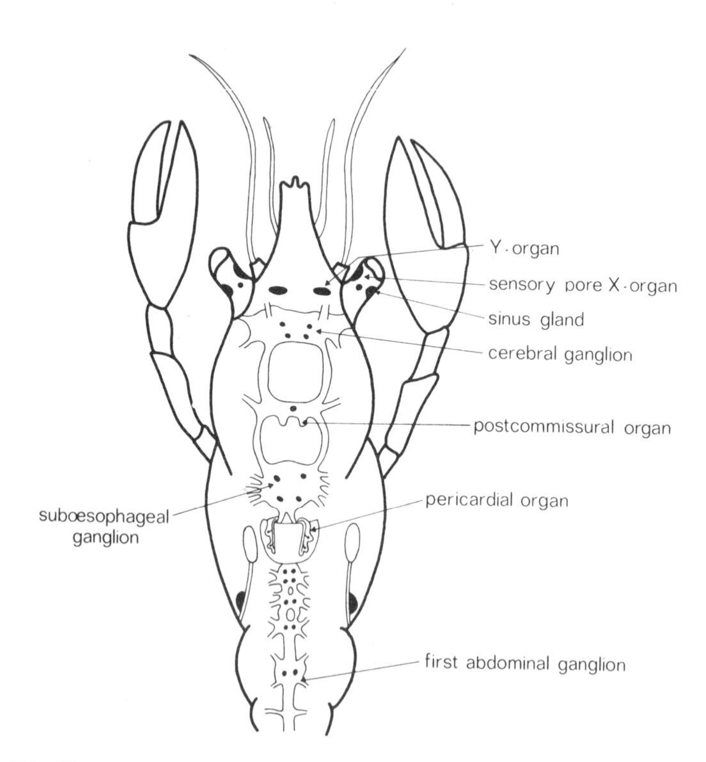

Figure 5.8 The neurosecretory system of a crustacean (after Gorbman & Bern, 1962, ref. 32).

would have no need of a blood-brain barrier. In fact there is, in decapod crustaceans at least, some restriction of access to the central nervous system from the main haemocoel.[19] Potassium ions, for example, penetrate into the nervous system relatively slowly. Substances dissolved in the intracerebral blood supply by contrast face no such barriers, and have free access to the nervous system. The reason for this arrangement may be that the intracerebral blood supply contains freshly oxygenated blood. It is possible also that it is relatively free of excretory products after passage to the excretory organs. Because of the known sensitivity of neurones to the "quality" of their environment, such an arrangement makes good sense physiologically. From the point of view of the neurosecretory cells in the central nervous system, however, the existence of the blood-brain barrier raises the problem of how to release their products into the general blood system. As in insects (p. 87), the solution has been the development of a set of neurohaemal organs. In the crustaceans which have been examined there are three main neurohaemal organs (figure 5.8).

The *sinus glands,* which lie in the eyestalks (or anterolateral to the brain in sessile-eyed forms), contain the endings of axons from cell bodies in the optic ganglia.

The *postcommissural organs* lie behind the tritocerebral commissure and contain axon endings from cell bodies in the brain. These organs are analogous in position to the corpora cardiaca of insects.

The *pericardial organs,* as their name indicates, lie close to the heart. They contain endings of neurosecretory cell bodies from several (at least) of the thoracic ganglia.

In addition to the neurosecretory cell bodies which lie in the various ganglia of the central nervous system, it has recently been discovered in the lobster that there are a considerable number of secretory cell bodies lying on the nerves leaving each of the thoracic ganglia.[20] The cells lie close to the point where the major thoracic nerves divide into their two main branches (figure 5.9). By injecting the cell bodies with the dye Procion Yellow they can be shown to have processes which branch and end near the cell bodies. Electronmicroscopy shows that any material released from these endings would have free access to the thoracic haemocoel. As well as these pericellular processes, the cells have axons which run to the pericardial organs lying in the pericardial sinus. Thus one has the unusual situation that neurosecretory material is released by these cells at two different sites (figure 5.9); one into the haemolymph before it enters the gills, and one into haemolymph leaving the gills. The significance of this arrangement is unknown.

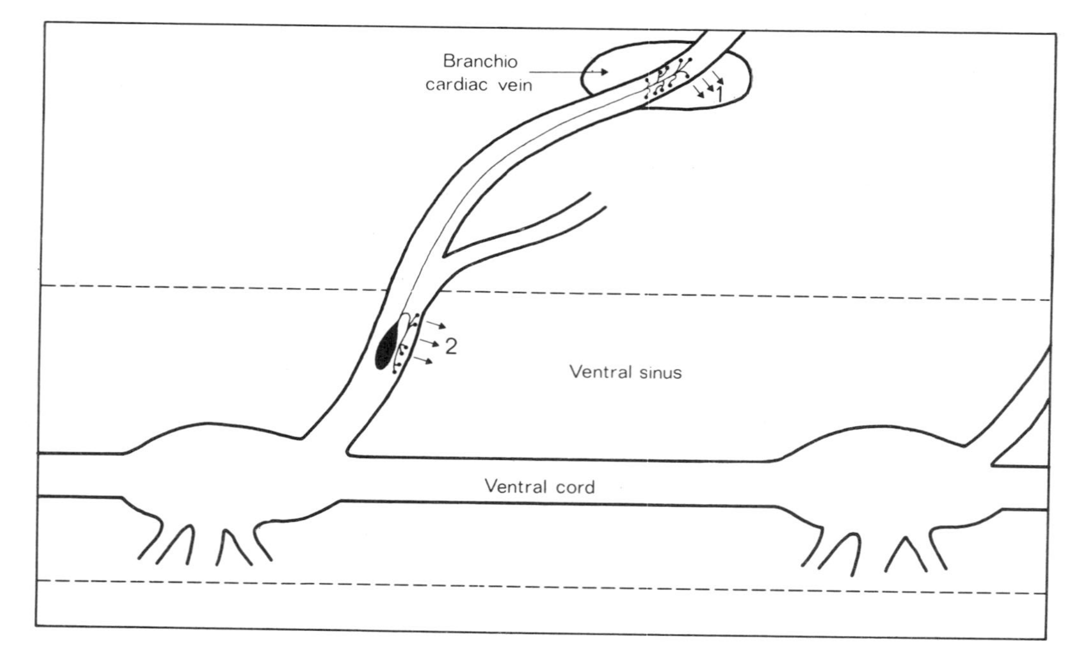

Figure 5.9 Representation of a root neurosecretory cell in the thorax of the lobster *Homarus americana*, showing the two release sites. At 1, octopamine is released into circulation after the gills, while at 2 release is into haemolymph before it passes to the gills (after Evans *et al.*, 1976, ref. 20)

Mollusca

Neurosecretory cell bodies have been described in all the different molluscan groups, though complete neurosecretory systems (cell bodies, axon tracts and release sites) have been demonstrated only in gastropods and cephalopods. The general features of the organization of these systems are illustrated in figure 5.10. The cell bodies are comparatively large (as

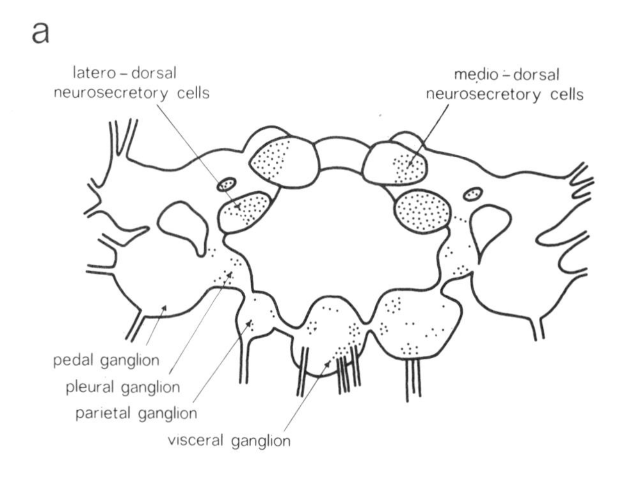

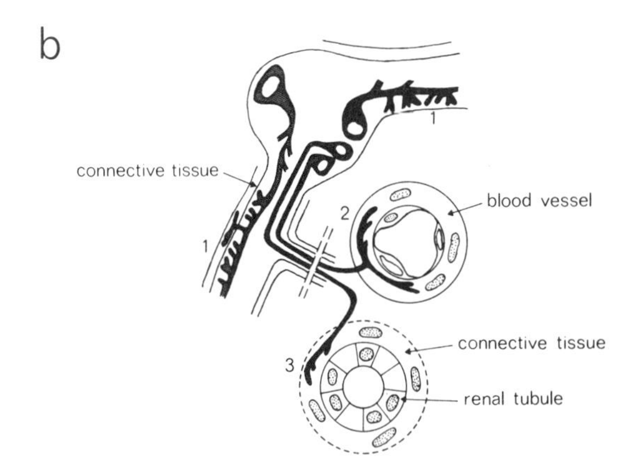

Figure 5.10 (*a*) The distribution of neurosecretory cells (black dots) in the central ganglia of a snail. (*b*) The release sites of molluscan neurosecretory cells: 1, at the periphery of ganglia and connectives; 2, in the walls of blood vessels; 3, in connective tissue surrounding other organs (after Joosse, 1976, ref. 22).

we have seen earlier (p. 33), many of them are very large indeed (up to 500 μm in diameter) and lie in relatively superficial positions in the ganglia. They thus make excellent material for investigations of electrical activity as we noted in chapter 3. As one would expect, the neurosecretory axons branch repeatedly; about 800 axon endings were found in one neurosecretory cell in *Lymnaea stagnalis*.[21]

As far as is known, molluscs have no blood-brain barrier. It is significant then that, in marked contrast to the situation in insects, crustaceans and vertebrates, the neurosecretory cells of molluscs do not release their neurohormones at definite neurohaemal organs. Instead, the release sites are widely distributed and extensive in nature, occurring on the surface of nerves, commissures and connectives, in the walls of blood vessels, or even on the surfaces of such organs as the kidneys.[22] From the argument that we advanced earlier that there might be a necessary correlation between the existence of a blood-brain barrier and the development of neurohaemal organs at some short distance away from the ganglia, the difference between molluscs and, say, insects, is just what one would predict.

Echinoderms

Following the discovery that extracts of the radial nerves of starfish will induce shedding of gametes when injected into a ripe starfish, attempts have been made to localize neurosecretory cells in these nerves. It turns out that there are numerous bipolar and multipolar ganglion cells, both in the nerve ring and the radial nerves, which react positively to neurosecretory stains.[23] In addition, there are so-called *supporting* cells peripherally located in the radial nerves which stain very intensely with paraldehyde fuchsin. The layer containing the supporting cells has been shown to be very rich in the peptide causing the gonads to release their products. Their distribution over a large area makes sense in an animal with a large extracellular fluid space, the contents of which are not rapidly circulated.

General features of invertebrate neurosecretory systems

Before we go on to consider the organization of the neurosecretory systems of vertebrates, it is worth considering the general points which can be made from our survey of invertebrate systems. Indeed, if they are general points, we ought to be able to use them to predict how vertebrate systems might be organized.

Probably the clearest point to emerge is that where the nervous system is isolated or protected to any extent from contact with the general extracellular space or blood system, the axons of neurosecretory cells in the central nervous system come to run together as neurosecretory tracts before crossing the blood-brain barrier at a very limited number of points. Only then do they branch repeatedly and achieve effective contact with the extracellular fluid into which their neurohormones are released. The numerous endings in such cases usually occur as definite swollen structures known as *neurohaemal organs*. The comparison of the insect and molluscan system is very revealing in this respect.

Where there is no or limited circulation of the extracellular fluid, the neurosecretory cells are relatively widely distributed; *Hydra* is a good example of this. Presumably this reduces diffusion delays between the time of release and the effect of the hormone.

In the larger invertebrates, the neurosecretory axons branch repeatedly before ending; while in the smaller invertebrates the cell bodies may be evident, but their endings are not. It seems very likely that this reflects the automatic consequence of increasing size that surface area to volume ratios decrease. If the neurosecretory cells in larger animals are to be able to achieve as high a concentration of their products in the extracellular space, they must release extra material; as a first step they seem to achieve this by huge increases in the number of endings, so that a single cell may have in the order of 1000 endings. In addition to this, the number and size of the neurosecretory cell bodies increases; insects of different sizes show this particularly clearly.

Finally, and appropriately enough, the neurosecretory axon endings lie close to the extracellular fluid that transports the hormone after release. This is particularly evident in the annelids. Here there is clear evidence that hormones can act if introduced into the coelomic fluid, and neurosecretory release sites are found which lie close to the coelomic cavity.

Neurosecretory systems in vertebrates

The neurosecretory systems of vertebrates, as one might expect, have been studied in much more detail than those of most invertebrates. Also, compared with the invertebrates, they are a much more homogeneous group. Accordingly we shall treat them in rather a different manner, and take in turn each of the different neurosecretory systems such as the neurohypophysis, the pineal organ, and so on, rather than consider each of the different vertebrate groups separately. With the exception of the lampreys and hagfishes (the cyclostomes), vertebrates have an effective blood-brain barrier. On the basis of what we have seen in invertebrates, we should expect to find, therefore, that the products of the neurosecretory cells of the central nervous system in most vertebrates released at neurohaemal organs. However, the release sites of neurohormones in cyclostomes might well be at the surface of the central nervous system.

The neurohypophysis[24]

As with all the invertebrate animals we have considered, the brain in vertebrates contains neurosecretory cells. Many of these cells lie in the hypothalamus and, because of the blood-brain barrier, these cells require a neurohaemal system at which to release their active products into the general blood circulation. The neurohaemal area concerned is called the *neurohypophysis* (see page 101). This is an area or projection from the floor of the brain, which is specialized in structure and richly vascularised to facilitate the release process. The entire system, i.e. neurosecretory cell bodies in the hypothalamus, axons running to the neurohypophysis and the neurohypophysis itself is known as the *hypothalamo-neurohypophysial system.* In higher vertebrates, the anatomy of the system is complicated by the fact that the glandular adenohypophysis and pars intermedia come to be in close association with the neurohypophysis, together forming the *pituitary gland.* The adenohypophysis and pars intermedia do not originate from the nervous system, but derive instead from the roof of the mouth. They are also concerned with hormone synthesis and release, though the cells concerned are not neurosecretory but glandular, and the hormones produced are rather different in structure from the amines or small peptides which neurosecretory cells typically produce. This situation of an association of a neurohaemal organ with a glandular structure derived from an adjacent epithelial or epidermal structure is not unique to vertebrates. The corpus cardiacum/corpus allatum complex of insects (p. 90) is a case in point. It is easy to imagine that, during evolution, a tissue lying close to a neurohaemal area may have become increasingly sensitive to the released neurosecretion, and that this increasing responsiveness may have led to the development of a gland regulated from the brain.

The main evolutionary trends in the anatomy of the hypothalamo-neurohypophysial system are:

(i) The neurosecretory cell bodies of the hypothalamus become aggregated into more definite areas. In lower vertebrates the cell bodies lie throughout the preoptic nucleus of the brain between the optic chiasma and the midbrain, but in the more highly evolved vertebrates, the cell bodies are confined to supraoptic and paraventricular nuclei which

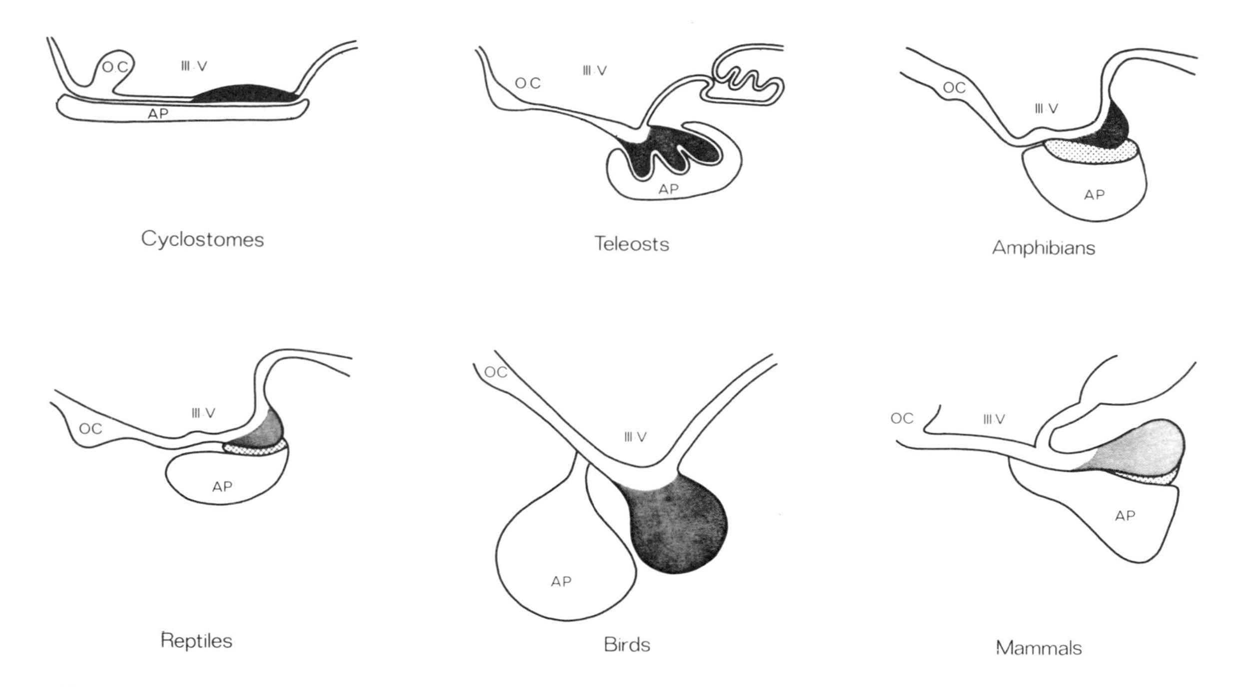

Figure 5.11 The organization of the pituitary gland in the different vertebrate groups. In each case the neurohypophysis is drawn densely stippled, while the pars intermedia is cross-hatched. OC—optic chiasma; III. V—third ventricle; AP—adenohypophysis.

lie respectively immediately behind the optic chiasma and in the median hypothalamus.

(ii) The floor of the midbrain, which constitutes the neurohaemal area, increasingly becomes protruded, eventually leading to a structure called the *neural lobe*, which is attached to the brain only by a slender stalk composed almost entirely of neurosecretory axons.

This latter trend is summarized in diagrammatic form in figure 5.11. Compared with, say, the pineal organ, the neurohypophysis is much more uniform in structure and function in the various vertebrates. It is, therefore, worth selecting for more detailed description only examples from the most primitive and most advanced vertebrates to illustrate the range of structure. The function of the system in terms of its control of various target organs is dealt with in the next chapter (p. 123).

The organization of the hypothalamo-neurohypophysial tract in cyclostomes (lampreys and hagfishes) is thought to differ little from the primitive condition. As figure 5.12 shows, the neurosecretory cell bodies whose axons run into the neurohypophysis are found throughout the preoptic nucleus. Most of their axons run along the ventral edge of the brain to reach the neurohypophysis; some, however, approach from a more lateral direction. In form, the neurohypophysis which contains the axon endings is little more than a thickened region of the floor of the brain. Unlike the arrangement in higher vertebrates, it does not project downwards beneath the general ventral surface of the brain. This may well be a reflection of the lack of a blood-brain barrier in these animals. In the neurohypophysis, the neurosecretory axons end at or near the many small blood vessels which run through it.

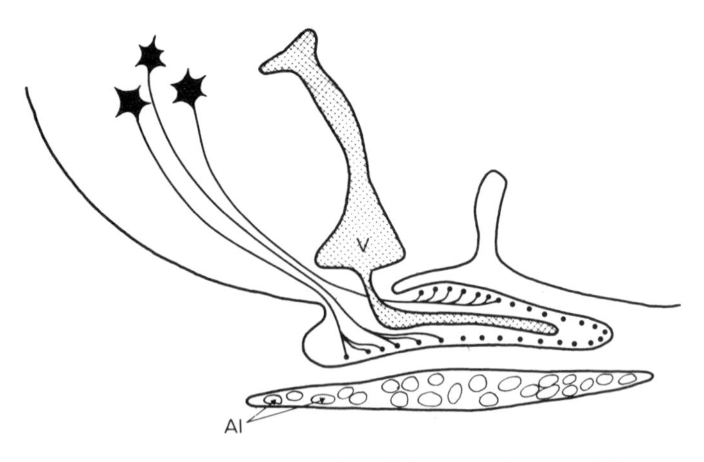

Figure 5.12 The organization of the hypothalamo-neurohypophysial system in a cyclostome. AI—adenohypophysial islets; V—third ventricle. The dots represent the neurosecretory nerve endings in the neurohypophysis (after Olsson, 1959, ref. 27).

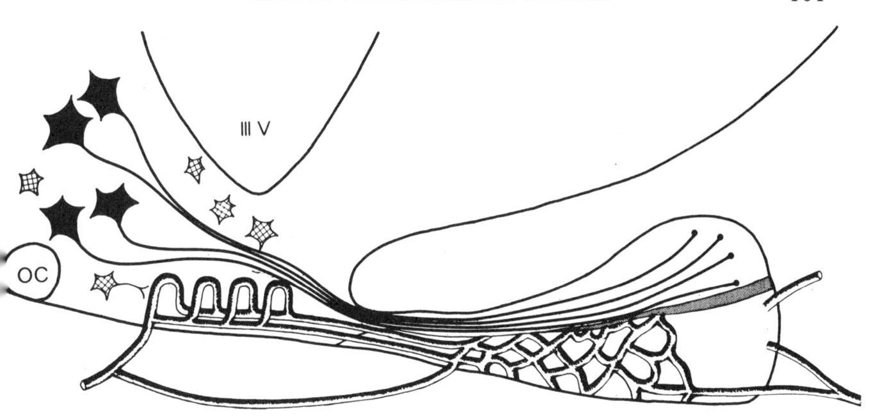

Figure 5.13 The organization of the hypothalamo- neurohypophysial tract in mammals. The neurosecretory cell bodies of the tract are shown in black; the other cell bodies shown (cross-hatched) are those of the parvocellular system, and they release products into the capillaries at the floor of the brain (lower left). OC—optic chiasma, III V—third ventricle.

Figure 5.13 illustrates the organization of the hypothalamo-neurohypophysial tract in mammals. In reptiles and birds it is organized essentially in a structurally similar fashion, although in birds the neurohypophysis is more clearly separated from the adenohypophysis (figure 5.11). The neurosecretory cell bodies lie in the supraoptic and paraventricular nuclei in the hypothalamus. Their axons run down through the infundibular stalk into the neurohypophysis, where they branch repeatedly and end in close proximity to the network of capillary vessels (figure 5.14).

Neural structures generally contain glial cells as well as neurones. As figure 5.14 and 5.15 show, the neurohypophysis is no exception; 30% of its volume is occupied by glial cells known as *pituicytes*. At one time it was supposed that the pituicytes were responsible for the secretion of oxytocin and vasopressin, the hormones released from the neurohypophysis. This was soon shown not to be the case since pituicytes in culture do not produce these hormones, and release of oxytocin and vasopressin continues for some time, although at a reduced rate, in neurohypophysectomized animals. The role(s) of the pituicytes remain(s) unclear. It is possible that they are trophic and supportive, but it has also been suggested that they may be concerned with the disposal of waste material, or with the local regulation of the concentration of potassium ions in the extracellular fluid, or be concerned in some way with feedback control of hormone release. There is as yet no good evidence to allow us to decide between these alternatives.

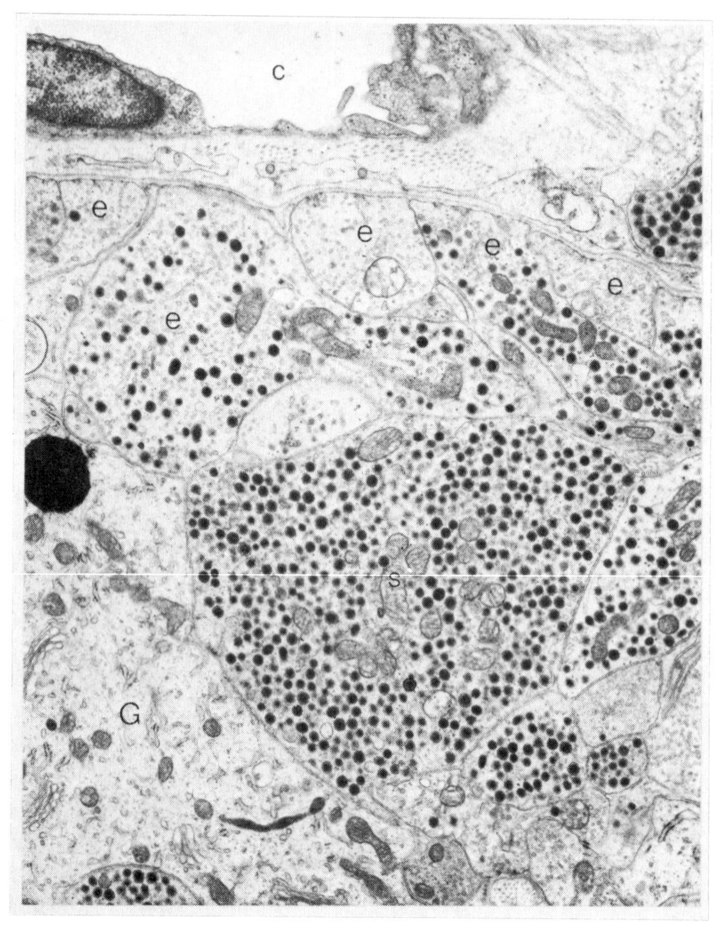

Figure 5.14 Electronmicrograph of the neurohypophysis of the rat to show the numerous neurosecretory axon endings (e) lying close to a capillary (c). G—glial cell (pituicyte); S—preterminal swelling of a neurosecretory axon. (Micrograph courtesy of J. F. Morris. × 5600)

The neurohypophysis as a neurohaemal organ

Before we leave a consideration of the neurohypophysis, it is worth emphasizing those of its features which contribute to its efficiency as a neurohaemal organ. The first obvious feature is that the axon terminals lie outside the brain at the end of the infundibular stalk (figure 5.13). Since they release their products into the blood, their contact with it has to be an intimate one, yet most of the brain is maintained in an environment kept separate from the blood by the blood-brain barrier. Having the terminals collected together and allowing them contact with the blood only at some distance from the brain ensures that the effect of this necessary gap in the blood-brain barrier is minimized. As we have seen earlier, just such a

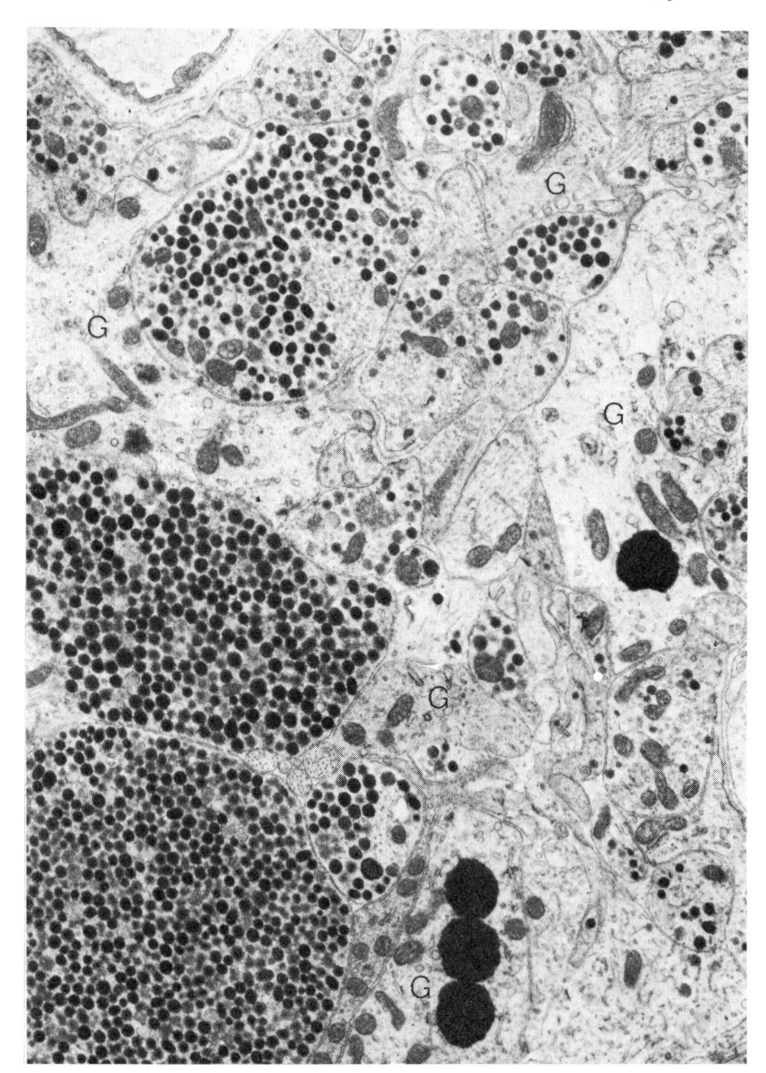

Figure 5.15 Electronmicrograph of the neurohypophysis of the rat to show the numerous glial cells (G) among the neurosecretory axon endings and swellings. (Micrograph courtesy of J. F. Morris. × 8400)

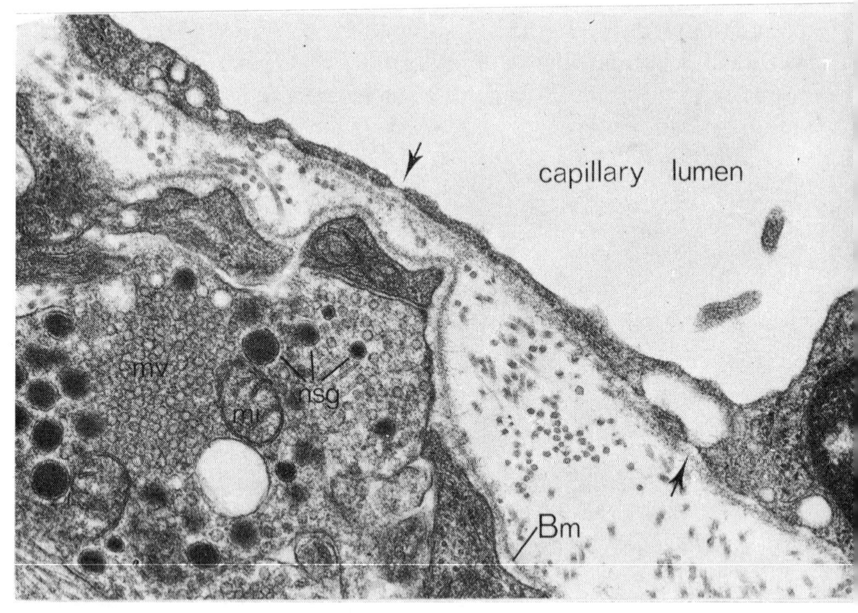

Figure 5.16 Electronmicrograph of the neurohypophysis of the rat showing the fenestrated epithelium (arrows) lining the blood capillaries. Bm—basement membrane; mi—mitochondrion; mv—microvesicles; nsg—neurosecretory granules. × 16 000.

solution is adopted by invertebrates with the same problem (see p. 83 for example).

If the axon terminals are to be able to liberate hormone into the blood at a high rate, it is important that the surface area of the terminals be large. This is achieved by very extensive branching of the axons once they have emerged from beneath the blood-brain barrier (figure 5.14). In the rat, for example, around 18 000 neurosecretory cells in the hypothalamus[25] release their products at about 40 000 000 axon endings in the neurohypophysis,[26] i.e. each cell has, on average, close to 2000 axon endings.

The final feature which is worth emphasizing is that the blood vessels at the site of neurohaemal contact are lined with a fenestrated epithelium rather than a continuous one (figure 5.16), and this must obviously contribute to the effectiveness of the system.

The pineal organ[28, 29]

The pineal organ in all vertebrates originates as an outgrowth of the dorsal surface of the diencephalic region of the brain. In lower vertebrates such as fish, amphibia, and reptiles, the organ either grows out through an aperture in the skull, or the area of the skull over it is thin and transparent. This allows the cells of the pineal organ to receive light directly. The difficulties in understanding the function of the pineal organ come from the fact that it has undergone dramatic changes in evolution. It seems certain that primitively it acted as a fully functional third eye. In higher vertebrates, however, it obviously can no longer act in this way as there is no route for light to reach it through the skin and skull, which are thick and opaque. As we shall see, in these animals the organ behaves as a neurohormone source and, while the effective stimulus may still be light, it is received through the eyes and transmitted as nerve impulses, via the brain, through the sympathetic nervous supply to the pineal organ.

Although the pineal organ may well have evolved originally from a structure whose only function was that of photoreception, it is now clear that this organ (even in living primitive vertebrates) acts also as a neurohormone source. Amongst living vertebrates, then, the pineal of the so-called lower forms functions both in photoreception and in the release of neurohormone(s), while in the more highly evolved vertebrates the photoreceptive function is lost and the organ functions only as a neurohormone source.

As a broad generalization, it may be said that the pineal organ of vertebrates usually contains the following elements. Of central importance there are pinealocytes, the primitive photoreceptors. These cells may make synapses of two types with nerve axons; there are afferent synapses with ganglion cells (sensory nerve cells), the axons of which are thought to run down into the brain, and there may be efferent synapses with axons from the sympathetic nervous system. The pinealocytes also contain in their basal processes small numbers of dense-cored vesicles which, as we shall see, are the ultrastructural indications of the ability of these cells to release neurohormones.

We shall now look at the pineal organs of representatives of the different vertebrate groups in order to illustrate just how much its structure and function has altered during evolution.

Cyclostomes

The earliest vertebrates, the Agnatha, as their name indicates had no jaws. They are known mostly from fossil forms, but there is one group, the Cyclostomes, which has survived to the present day and which contains the lampreys and hagfishes. The pineal of lampreys is worth consideration in that these animals are thought to show some of the characteristics of the earliest vertebrates. Strangely, hagfishes have lost all trace of the pineal system; this ties in with the widely held idea that lampreys and hagfishes are not closely related.

In lampreys there is, originating from the roof of the diencephalon, the so-called third or median eye. It consists of a pair of unequally developed sacs, each with a narrow lumen; the right-hand one, the pineal, is larger and lies above the smaller left parapineal (figure 5.17). Figure 5.18 shows the structure of the typical cells found in the lamprey pineal organ. The walls of the sacs contain photosensory cells and also ganglion cells which are thought to synapse with the sensory cells and send axons down into the brain. There are also pigment and supporting cells present. The photosensory cells have apical structures protruding into the lumen of the sacs; these protruding apices have an ultrastructure very similar to that of the outer segments of the rods and cones of the retina. They contain a swollen outer part packed with flat stacks of membranes, and a slender cilium-like process connecting the outer part with the main cell body. The connecting stalk contains characteristic ciliary microtubules with nine

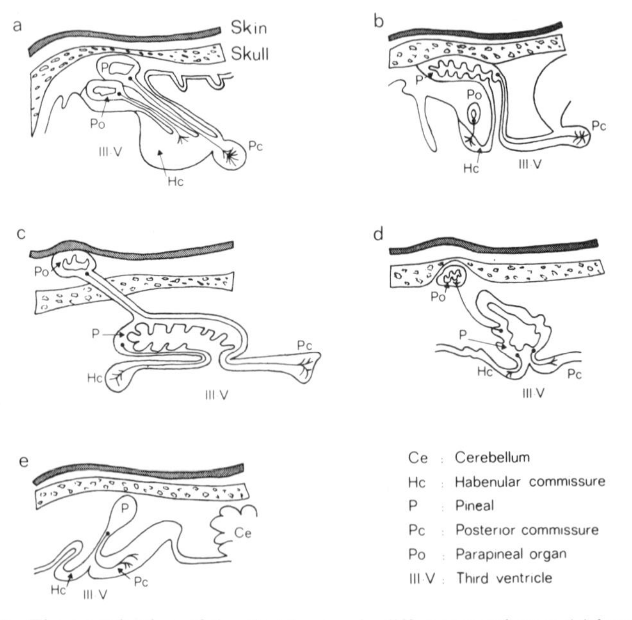

Figure 5.17 The organization of the pineal organ in different vertebrates. (*a*) lamprey, (*b*) teleost fish, (*c*) amphibian, (*d*) reptile, (*e*) bird (after Collin, ref. 28).

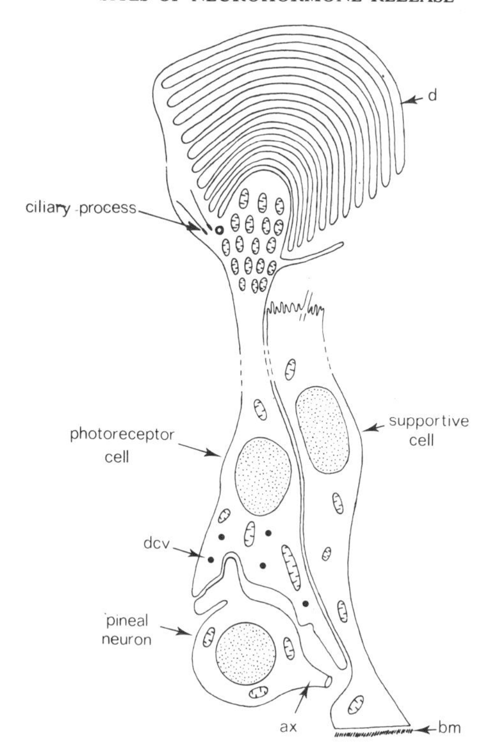

Figure 5.18 The structure of the typical cells found in the lamprey pineal organ. ax—axon; bm—basement membrane, d—distal processes, dcv—dense-cored vesicles (after Collin, ref. 28).

peripheral doublets but, as in retinal photoreceptors, it lacks the central pair found in most cilia and flagella. The fibrils end in an axial centriole at right angles to a second centriole nearby.

In their inner segments the photosensory cells contain vesicles of two sorts: small (30–50 nm) vesicles in association with presynaptic ribbons, and larger (up to 300 nm in diameter) dense-cored vesicles more randomly distributed in the cytoplasm. It is thought that the sensory cells make synapses with the ganglion cells and that the smaller vesicles contain the transmitter used at these synapses. The nature of the contents of the dense-cored vesicles is not known, but it is suggested that they may contain melatonin which is released into circulation and may act to control pigmentation (see below).

Rather little is known about the function of the pineal eyes of lampreys. However, they appear to be active in controlling the daily rhythm of colour change which the ammocoete larva shows. If the pineal eyes are removed, the animals remain continually dark in colour and no longer become pale at night as do control animals.

How this control is effected is not clear. The organ may act solely as a photoreceptor and pass information to the brain, or it may itself release melatonin, a blood-brain factor known to cause pigmentation changes. Melatonin is found in the pineal organs of many animals; it acts to cause profound changes in the density of pigmentation of chromatophores.

In summary then, the pineal organ of cyclostomes is an active photoreceptor which may, in addition to passing information to the brain, produce neurohormone(s) in response to changes in illumination.

Fish

Among fish, the pineal organ has been investigated in Chondrichthyes (which includes sharks and rays) as well as in Osteichthyes (which include most bony fish). In these animals, the organ is rather similar to that of cyclostomes, and there is again a small so-called parietal organ which is probably homologous to the smaller sac (parapineal) found in cyclostomes (figure 5.17). The photosensory cells again have an ultrastructure similar to that of rods or cones; they make synaptic contacts with ganglion cells whose axons run to the brain, and they contain dense-cored vesicles. As in the lamprey, it is supposed that the dense-cored vesicles contain melatonin. This substance is found in the pineal organs of some fish and, in the pike, the pineal contains large amounts of 5-hydroxytryptamine (5-HT) which is the precursor for melatonin synthesis (figure 5.19). Until recently it was thought that the pineal organ of the fish did not receive an efferent nervous supply, but it now seems that in at least in one species of fish there are a few sympathetic fibres supplying the pineal.

Very probably then, in fish, the pineal organ operates much as in cyclostomes, as both a light sensor and a source of neurohormone.

Amphibia

In Amphibia, the pineal organ differs from those in cyclostomes and most fish in having a definite efferent nervous supply from the sympathetic nervous system. These autonomic fibres contain catecholamines, and it is thought that they might exert a controlling influence on the pineal and its secretory processes. Both the enzyme, hydroxyindole-O-methyltransferase

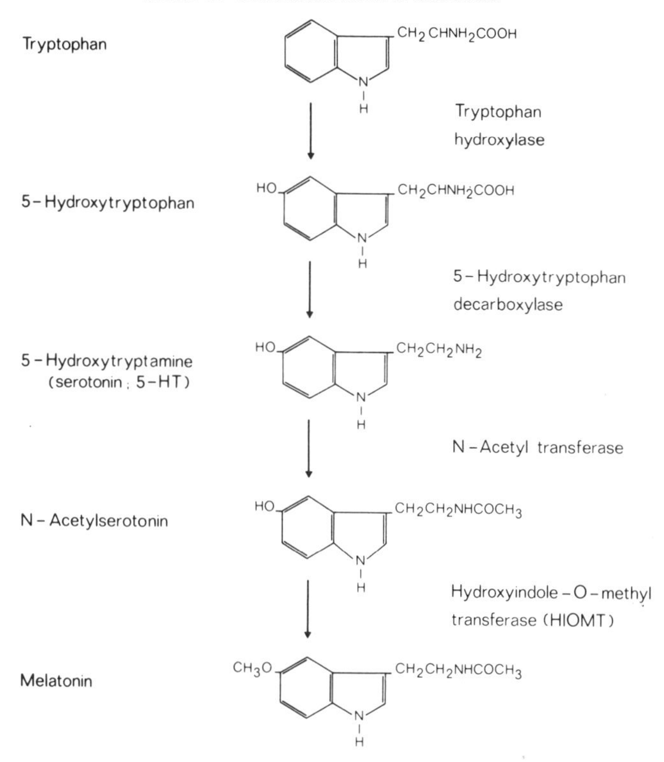

Figure 5.19 The biochemical steps involved in the synthesis of melatonin from tryptophan

(HIOMT), which is necessary for the synthesis of melatonin from 5-HT, and 5-HT itself have been found in the pineals of amphibians; so it looks very likely that melatonin is synthesized from 5-HT in this organ. It is not actually known in which cells melatonin synthesis goes on, but it is suggested that the photosensory cells are responsible, and that melatonin is contained in the dense-cored vesicles that occur in these cells.

At least in urodele amphibians, the pineal cells have a well-developed photoreceptive apparatus of membrane stacks in the outer segments of the cells, and these are connected by the usual ciliary stalk to the inner parts of the cells. The inner parts contain synaptic vesicles and ribbons, and the cells synapse with ganglion cells; and it is known that the system responds by sending nerve impulses to the brain when the organ is illuminated.

Other amphibians have yet to be examined; they may well show a range of pineal structure such as is found in the different reptile groups.

Reptiles

Among reptiles, the pineal gland shows a wide diversity; it is absent in crocodiles, while in some lizards it is well developed and may still retain a photosensory function; and in snakes it seems only to have a secretory function. Broadly speaking, however, in most reptiles such as Lacertilia, which group includes lizards, and particularly in Chelonia, the tortoises and turtles, most of the pineal cells show only a rudimentary development of the photosensitive apparatus. These rudimentary photoreceptors make very few synaptic contacts with dendrites from the few remaining sensory nerve cells, and there is a corresponding reduction in the number of sensory axons running down into the brain. Strangely, however, among the large numbers of rudimentary photoreceptors there are usually to be found a few pineal cells which still retain a fully developed photosensory system. That the role of the pineal organ is largely one of secretion is supported by the finding that 5-HT can be extracted from the gland and that histochemically it is most evident in those areas of the pineal cells where the dense-cored vesicles occur in the greatest numbers. The pineal organs of lizards contain considerable amounts of the enzyme HIOMT, suggesting that melatonin is synthesized from the 5-HT. Particularly in turtles, autonomic nerve fibres can be seen to supply the organ, and histochemical studies show them to contain catecholamine, probably noradrenaline.

Birds

The pineal organ of birds is rather similar to that of reptiles, but the photosensory apparatus has regressed even more and, except in a few cases, the pineal cells appear to make no synapses with sensory nerve cells. There is a well-developed supply of nerve fibres from the sympathetic system, at least some of them containing noradrenaline. The rudimentary photoreceptor cells contain many dense-cored vesicles originating from the Golgi zone, and again they are thought to contain melatonin. As in amphibians and some reptiles, the pineal organ contains 5-HT and HIOMT, suggesting that melatonin is synthesized from 5-HT. In some species of birds there are still present at least the cell bodies of the sensory nerve cells (ganglion cells) as a vestigial remnant of the primitive photosensitive function.

Mammals

In mammals, the pineal organ is exclusively secretory. The pineal cells now have lost almost all trace of the external segment, except for a ring of nine ciliary microtubules at the distal ends of the cells. Such cells are termed *pinealocytes*. They are innervated only by autonomic nerve fibres from the superior cervical ganglia. This sympathetic innervation is thought to depend on noradrenaline as a transmitter. The pinealocytes have short processes which end on the basement membrane of the pineal organ, or at intercellular spaces which communicate with the perivascular spaces. These perivascular spaces may extend well into the organ, presumably to increase the ease with which active compounds released in the organ can reach the blood supply. Just such an arrangement is, of course, characteristic of endocrine organs.

As one would expect, melatonin, its precursor, and enzymes specific for its biosynthesis have all been isolated from mammalian pineal organs. In addition, however, there are claims that the pineals of mammals secrete peptides which affect the development of the gonads.

Summary

The pineal organs of vertebrates vary widely in their organization. The lower vertebrates are thought to show what may have been the primitive condition in which the pineal organs act both as photoreceptors and as endocrine organs. The pineal organs of mammals are exclusively endocrine, while intermediate conditions are encountered in reptiles and birds. In evolution, the structure of the characteristic pineal cell has altered from a fully developed photosensitive cell to a rudimentary photoreceptor cell, and finally to a pinealocyte (figure 5.20). These changes are accompanied by concomitant changes in the innervation of the organ. Those pineal organs which are directly photoreceptive send afferent fibres to the brain, while those that are not, receive efferent fibres from the sympathetic nervous system. The function of the pineal organ as a source of neurohormones is considered in the next chapter (p. 137).

Chromaffin tissue: suprarenal bodies, adrenal medulla[30]

In the early embryological development of vertebrates, the cells of the neural crest give rise, amongst other structures, to the nerve cells of the autonomic nervous system. Most of these embryonic cells develop into sympathetic and parasympathetic ganglion cells. However, others give rise to the so-called chromaffin cells with their characteristic staining properties. These cells synthesize, store and secrete catecholamines (such as dopamine, adrenaline and noradrenaline, figure 5.21). Because of their

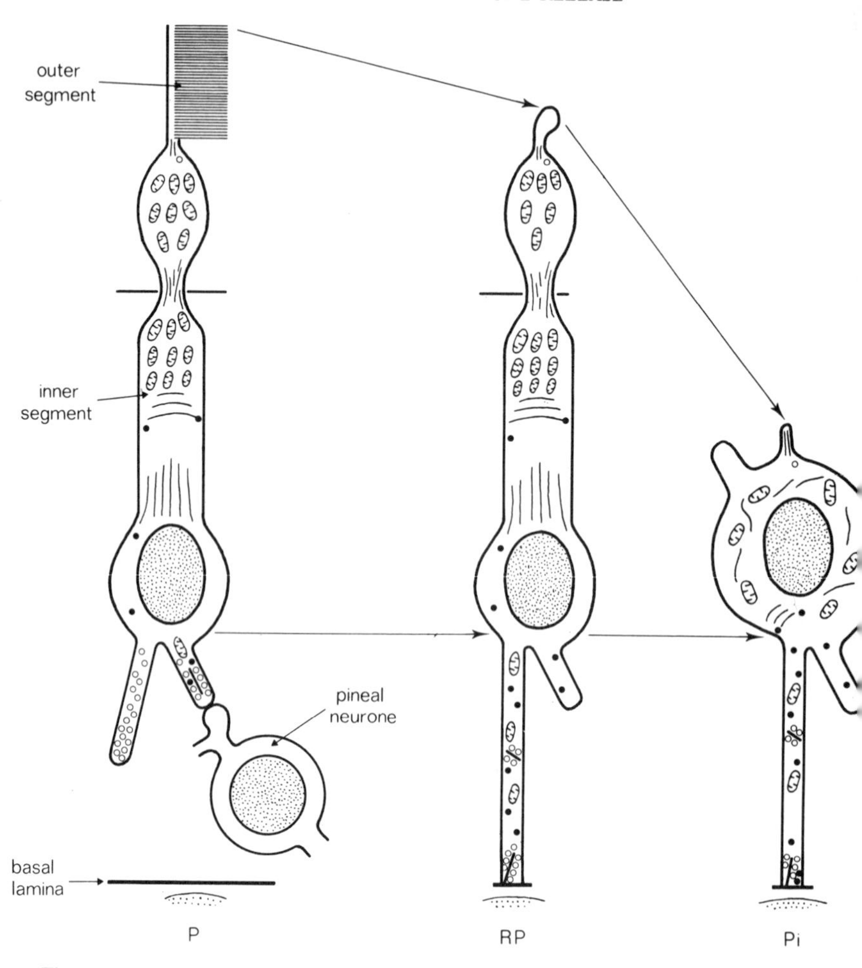

Figure 5.20 The phylogenetic transformation of the photoreceptor cell of the pineal organ. The photoreceptor cell (P) is thought to be similar to that occurring primitively. It is supposed that, during evolution, the photoreceptor cell has given way successively to a rudimentary photoreceptor cell (RP) and then to a pinealocyte (Pi) (after Collin, ref. 28).

catecholamine content, they give an intense yellow-brown reaction product when they are soaked in solutions of chromic acid or of chrome salts, hence the name *chromaffin cells*. Like ganglion cells of the sympathetic nervous system, they are innervated by preganglionic sympathetic nerve fibres. They can to some extent be regarded as the equivalent in the sympathetic nervous system of the neurosecretory cells in the central nervous system. In their evolution, their axons have been very much reduced and they release their active compounds into the general circulation, more or less directly from the surface of the cell body. In this method of hormone release from a relatively limited area, there is an apparent paradox. We have seen how the neurosecretory cells of the central nervous system produce a rapid release of their active compounds by an enormous increase in the area of their release sites. This they achieve by a proliferation of the axon endings to the extent that hypothalamic neurosecretory cells may each have around 2000 axon endings. Yet in each chromaffin cell of the mammalian adrenal medulla, for example, release is confined to a part only of the surface of the cell, and it is not elaborated in any way.

The explanation of this paradox seems to be that the chromaffin cells make up in number what they lack in specialization of release sites. For example, the hypothalamus of the rat contains about 18 000 neurosecretory cells responsible for the synthesis of two hormones (oxytocin and vasopressin). The neurohypophysis contains approximately 4×10^7 endings at which these hormones are released. The rat adrenal medulla, also responsible for the release of two hormones (adrenaline and noradrenaline), has a volume of about 2 μl. As an approximation, each

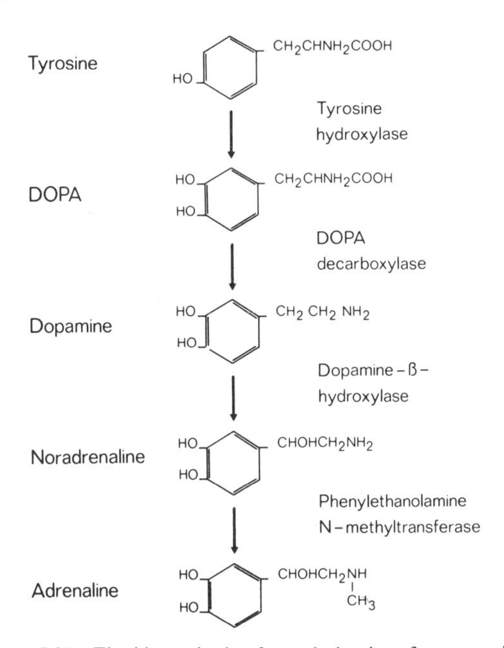

Figure 5.21 The biosynthesis of catecholamines from tyrosine.

chromaffin cell may be considered to occupy the space of a 20-μm cube. From this it follows that the adrenal medulla contains around $2{\cdot}5 \times 10^5$ cells. Since the surface area from which hormone is released is a good deal larger in a chromaffin cell than in a neurosecretory axon ending in the neurohypophysis, the hormone release sites in adrenal medulla and neurohypophysis are probably of rather similar areas.

Evolution and organization of chromaffin tissue

Cells which synthesize catecholamines are not confined to vertebrates; their presence in invertebrates has been known for some time. Molluscs, annelids, crustaceans and insects all have been shown to contain significant numbers of these cells. In nearly all cases, the cells occur in the ganglia of the central nervous system, either in the brain, or in the ganglia of the main nerve cord. They are, therefore, not homologous with the chromaffin cells with which we are chiefly concerned in this section, all of which form part of the autonomic nervous system. They may well also differ from the cells of the adrenal medulla, for example, in that they are likely to act as interneurones with their catecholamine products serving as neurotransmitters, never emerging into the circulation to act as hormones. It is relevant to add that dopamine-secreting nerve cells occur in the central nervous system of many vertebrates, and all the evidence is that their effects occur within the confines of the central nervous system.

In some invertebrates there are amine-secreting nerve cells which occur outside the central nervous system (p. 87 and p. 93). These cells either secrete phenolamines such as octopamine (but not catecholamines) or their axons run directly into the central nervous system—and so presumably release their products there and not into the general circulation. Even in the cells which synthesize phenolamines and release them into circulation as hormones, the number of cells is rather few and the release occurs at multiple axon endings. This is in contrast with vertebrate chromaffin cells where there are many cells, each releasing only from the surface of the cell body. In short, there appear to be no clear homologues or even analogues of the vertebrate chromaffin cell in the invertebrates.

In contrast, chromaffin cells are found throughout the vertebrates. Figure 5.22 shows the disposition of chromaffin tissue in the various vertebrate groups. In so far as one can read the differences as an evolutionary series (the difficulty, of course, being that we only have the structure of present-day forms to go on), it seems that an originally linearly strung-out series of small collections of cells have increasingly become

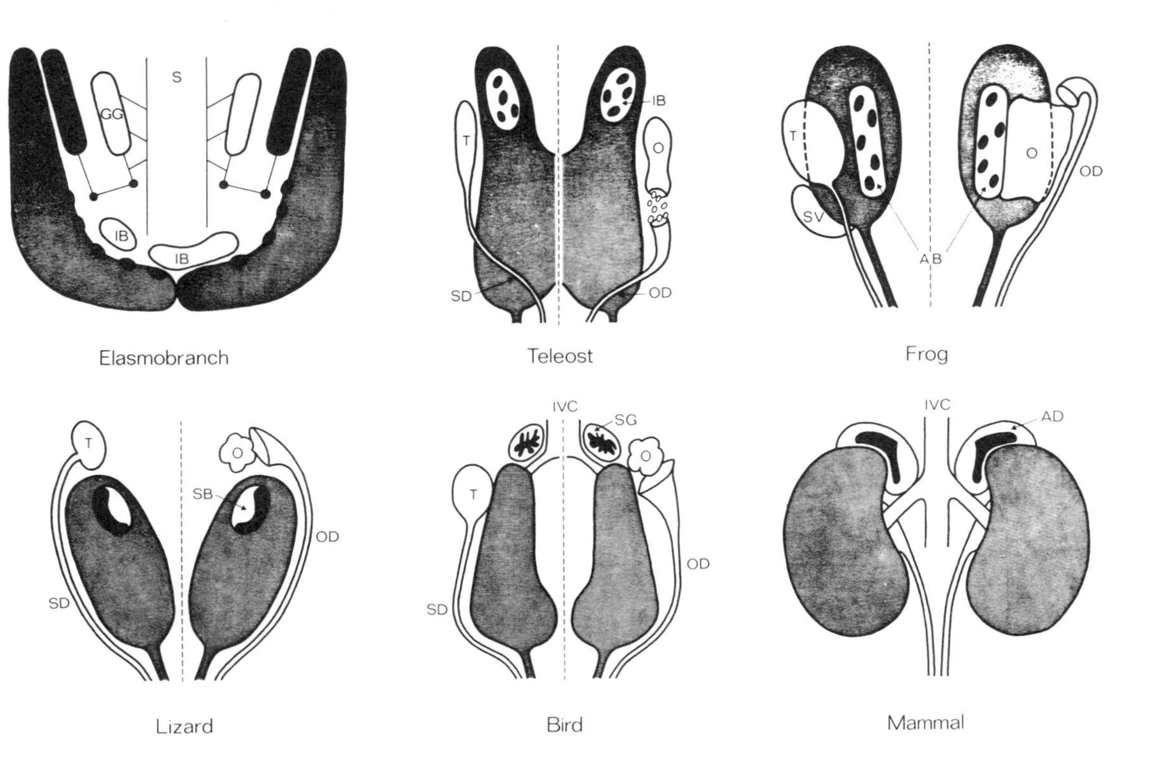

Figure 5.22 Chromaffin tissue in representatives of the different vertebrate groups. In each case, chromaffin tissue is shown black and renal tissue as stippled. AB, adrenal body; AD, adrenal gland; GG, gastric ganglion; IB interrenal body; IVC, inferior vena cava; O, ovary; OD oviduct; S, stomach; SB, suprarenal body; SD, sperm duct; SG, suprarenal gland; SV, seminal vesicle; T, testis.

aggregated into more compact bodies. Rather similar changes occur in the kidneys, and the chromaffin tissue often lies close to them or even embedded in them. The resulting organs are often, therefore, termed *suprarenal bodies* or, as in mammals and birds, *adrenal glands*.

Where the structure of these bodies or glands has been studied, it reveals an organization typical of neurohaemal organs involved in releasing quick-acting hormones. There is a rich capillary blood supply offering easy access to the chromaffin cells, and there are no enveloping layers of glial cells to slow the diffusion of released hormone into the blood vessels. The chromaffin cells are directly innervated by preganglionic sympathetic nerve fibres. These fibres stimulate the chromaffin cells to liberate their catecholamine hormones by acetylcholine released at the synapses between the fibres and the chromaffin cells.

The urophysis[31]

In several groups of fish, with the notable exception of the elasmobranchs, there exists a remarkable neurohaemal organ not seen in higher vertebrates. This is the urophysis which is a swollen structure found attached to the posterior end of the spinal cord (figure 5.23). Essentially, it consists of an association of many neurosecretory axon endings with an extensive capillary network. It also contains glial elements and connective tissue. The evidence for concluding that the urophysis is a neurohaemal organ is as follows.

The axon endings found in it contain very large numbers of electron-dense granules, and the axons can be traced back to large cell bodies in the spinal cord, called *Dahlgren cells*. The Dahlgren cells show all the ultrastructural features associated with neurosecretion; there is much evidence of granule formation by the Golgi complex, and there is well-developed endoplasmic reticulum. The convergence of the axons from these cells into a dense mass of branched axon endings which is richly vascularized is a very clear indication of the neurohaemal function of the urophysis. So, indeed, is the appearance of the basement membrane round the blood capillaries; it has many deep extensions which run in among the neurosecretory axon endings and which greatly increase the effective area of contact between the endings and the blood. In addition, electrical stimulation of the system in *Tilapia mossambica,* for example, causes the appearance in the axon endings of numerous electron-lucent vesicles. These, as we have seen (p. 75), very probably represent the results of the process of membrane retrieval after exocytosis.

Figure 5.23 Longitudinal section of the posterior end of the spinal cord of a fish (mullet) to show the swollen urophysis. (Micrograph courtesy of H. A. Bern. × 18).

In spite of what seems an impressive body of structural evidence strongly suggesting that the urophysis is a source of neurohormones, there is surprisingly little good physiological evidence that this is the case. Such evidence as there is indicates that the urophysis plays a role in osmoregulation and in the control of blood pressure. However, no purified hormones have been isolated. Unravelling the function of the system remains a major challenge to physiologists and neuroendocrinologists.

For the present we need to look at the organization of this caudal (tail) neurosecretory system in a variety of fish to see how the system has evolved and how it is organized to achieve its supposed function of releasing neurohormones.

Neurosecretory cells are found in the spinal cord over variable amounts of its length. In elasmobranchs, which lack a urophysis, the cells may be spread out over an area corresponding to at least the last fifty of the posterior vertebrae. In teleosts, however, the cells occur only in a length of cord corresponding to the last three to ten vertebrae (figure 5.24).

In elasmobranchs, the neurosecretory cells are not only well spread out. but the cell bodies are of remarkably large size, considerably larger than typical motor neurones. The axons are relatively short, running only as far as the surface of the cord, where the axons branch repeatedly. The endings terminate on areas of the meninx, where there is a close association with

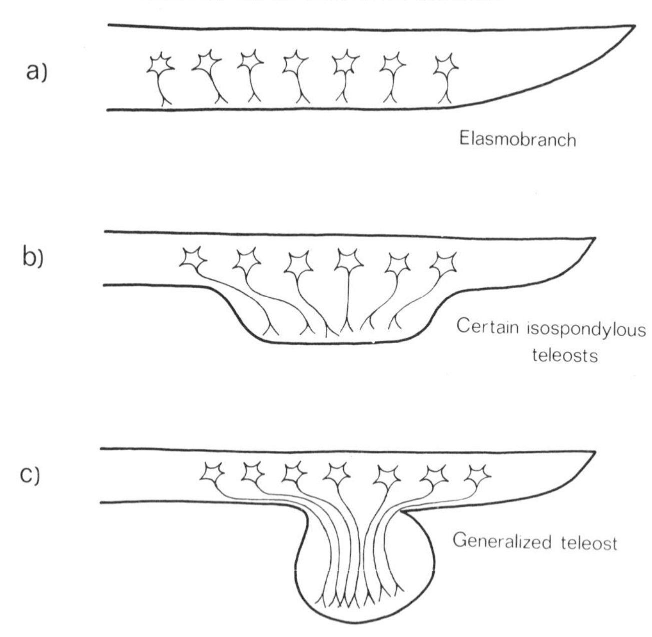

Figure 5.24 The organization of the urophysis in different fish (after Bern, 1969, ref. 31).

many finely branched blood vessels. The neurohaemal area is thus very extensive. That it is not more compact and removed a short distance from the spinal cord may reflect the way in which the blood composition is controlled in these animals. Unlike marine teleosts, which maintain the osmotic concentration of the blood well below that of the sea, the blood of marine elasmobranchs, although it contains just as low ionic concentrations, is made iso-osmotic with the sea-water by retention of large amounts of urea. As a result, it may well be that elasmobranchs are not in such danger as teleosts are of having to face large temporary changes in blood composition caused by damage or disease of the ion-regulatory mechanisms. In the teleost, such failure would swiftly lead to osmotic loss of water, which in turn would cause rapid changes in ion levels. By contrast, the speed of changes in ion levels in elasmobranch blood would be limited by the relatively slower process of ion diffusion into the blood from the sea. All this is to say that elasmobranchs may not need the elaborate machinery of a well-developed blood-brain barrier that teleosts would appear to need. The multiple penetration of the epithelia around the spinal cord by the neurosecretory axons of elasmobranchs may not, therefore, be as hazardous to the well-being of the neurones of the cord as might be the case in teleosts. Some support for this interpretation comes from hagfishes. In these animals, the problem of maintaining the main blood ions at constant levels has been solved by allowing them to rise in evolution until they match those of the sea. Appropriately, hagfishes have no blood-brain barrier at all.

In most teleosts, the axons running from the neurosecretory cells first come together as a tract before they leave the cord to end in a definite organ, the urophysis (figure 5.25). In teleosts such as the angler fish, where the urophysis is well separated from the spinal cord, the tract forms a stalk like that of the neurohypophysis (p. 98).

In some teleosts there are accessory neurohaemal structures of various

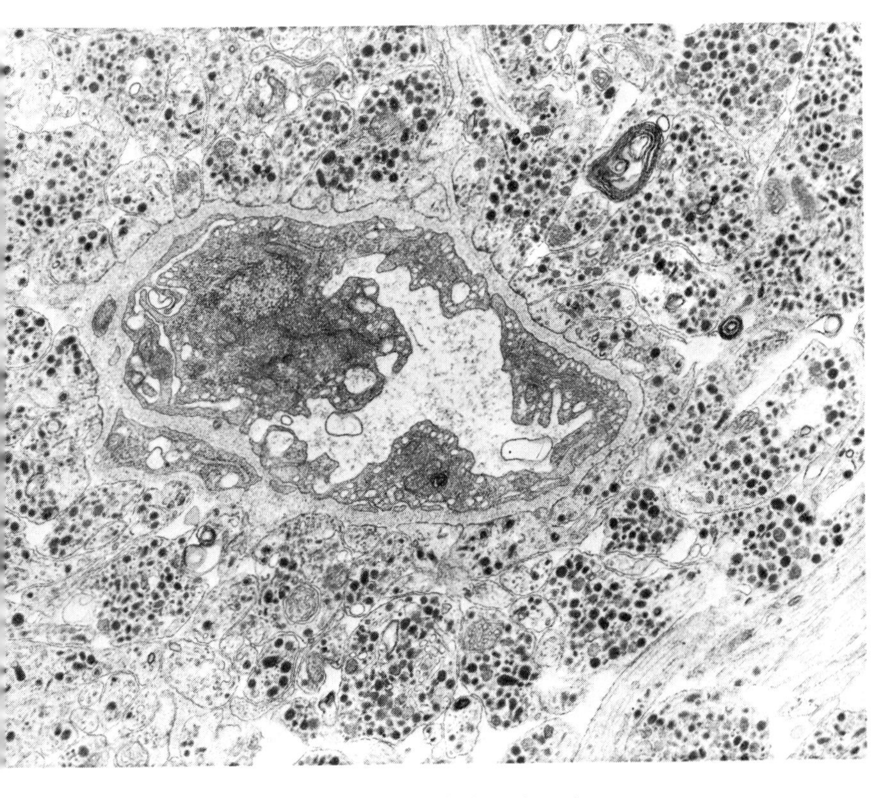

Figure 5.25 Electronmicrograph of a fish urophysis to show the many neurosecretory axon endings that cluster round a capillary branch (centre). (Micrograph courtesy of H. A. Bern. × 11 000)

kinds. In the flounder there is a small secondary urophysis, while in tunny fish there are multiple urophyses. In all cases, however, the neurosecretory axons end at some short distance from the nerve cord, just as they do in insects (see p. 87) where there is known to be an efficient functioning regulatory blood-brain barrier. As we have now seen in many cases, it is a general feature of neurohaemal organs that they are most distinct from the central nervous system in those animals which need the most effective blood-brain barriers.

Finally, it should be said that not only is there evidence of caudal neurosecretory systems in teleosts and elasmobranchs, but also in Holostei, Dipnoi and possibly also in cyclostomes. However, no traces of the system are to be found in the higher tetrapod vertebrates.

Production of peptides by cells from the neuroectoderm—the APUD concept[32]

A puzzling feature of recent research on physiologically active peptides in vertebrates has been that many of them are found not only in the brain, but also in various other cells, often in the gut and the pancreatic islets. A good example is somatostatin, a tetradecapeptide found in the hypothalamus, where its function is the local one of inhibiting the release of growth hormone from the anterior pituitary (adenohypophysis). This substance has now been detected in the stomach and pancreas, and in nerve fibres in the wall of the intestine in amounts equal to those occurring in the hypothalamus. It has also been found to inhibit the release of other endocrine peptides such as insulin, glucagon, and gastrin.

There is a simple concept that goes a long way to explain these peculiar findings. The idea is that cells producing peptide hormone are all derived from the neuroectoderm or from placodal ectoderm. Consistent with this, for example, is the finding, from embryological studies, that the hypothalamo-neurohypophysial complex is neuroectodermal in origin. Many of the peptide-producing cells have common cytochemical properties (the initial letters of which provide the acronym APUD by which the cell series are often known—these properties are a content of Amine, and an ability in Amine Precursor Uptake and Decarboxylation). Cells of the APUD series are found in the pituitary and pineal glands, in the hypothalamus, pancreas, stomach, intestine, thyroid gland, carotoid body, adrenal skin, lung, and urogenital system. What effects these peptides have, and how their actions are correlated with peptides from the central nervous system, are intriguing problems for the future.

REFERENCES

1. Abbott, N. J. & Treherne, J. E. (1977) "Homeostasis of the brain microenvironment: a comparative account." In *Transport of Ions and Water in Animals* (eds. B. L. Gupta, R. B. Moreton, J. L. Oschman & B. J. Wall), pp. 481–510, New York & London, Academic Press.
2. Treherne, J. E. (1974) "The environment and function of insect nerve cells." In *Insect Neurobiology* (ed. J. E. Treherne), pp. 187–244, Amsterdam & Oxford, North-Holland Publishing Co.
3. Gabe, M. (1966), *Neurosecretion,* New York, Pergamon Press.
4. Schaller, H. C. (1973) "Isolation and characterization of a low-molecular-weight substance activating head and bud formation in hydra," *J. Embryol. exp. Morph.* **29,** 27–38.
5. Lender, T. & Klein, N. (1961) "Mise en évidence de cellules sécrétices dans le cerveau de la Planaire *Polycelis nigra.* Variations de leur nombre au cours de la régeneration postérieure, "*C.r. hebd. Séanc. Acad. Sci. Paris* **253,** 331–333.
6. Silk, M. H. & Spence, I. M. (1969) "Ultrastructural studies of the blood fluke–*Schistosoma mansoni.* III. The nerve tissue and sensory structures," *S. Afr. J. med. Sci.* **34,** 93–104.
7. Davey, K. G. & Hominick, W. K. (1973) "Endocrine relationships between Nematodes and their Insect hosts–a review," *Experimental Parasitology* **33,** 212–225.
8. Baskin, D. G. (1976) "Neurosecretion and the endocrinology of nereid polychaetes," *Amer. Zool.* **16,** 107–124.
9. Prabhu, V. K. K. (1961), "The structure of the cerebral glands and connective bodies of *Jonespeltis splendidus* Verhoeff (Myriapoda: Diplopoda)," *Z. Zellforsch. mikrosk. Anat.* **54,** 717–733.
10. Maddrell, S. H. P. (1974), "Neurosecretion." In *Insect Neurobiology* (ed. J. E. Treherne), pp. 307–357, Amsterdam & Oxford, North-Holland Publishing Co.
11. Finlayson, L. H. & Osborne, M. P. (1968) "Peripheral neurosecretory cells in the stick insect (*Carausius morosus*) and the blowfly larva (*Phormia terrae-novae*)," *J. Insect Physiol.* **14,** 1793–1801.
12. Orchard, I. (1976) "Calcium dependent action potentials in a peripheral neurosecretory cell of the stick insect," *J. Comp. Physiol.* **112,** 95–102.
13. Orchard, I. & Osborne, M. P. (1977) "The effects of cations upon the action potentials recorded from neurohaemal tissue of the stick insect," *J. Comp. Physiol.* **118,** 1–12.
14. Raabe, M., Baudry, N. & Provansal, A. (1970) "Recherches sur l'ultrastructure des organes neurohémaux périsympathiques des Vespidae (Hyménoptères). Les organes latéraux longitudinaux," *C.r. hebd. Séanc. Acad. Sci, Paris* **271,** 1210–1213.
15. Brady, J. & Maddrell, S. H. P. (1967) "Neurohaemal organs in the medial nervous system of insects," *Z. Zellforsch. mikrosk. Anat.* **76,** 389–404.
16. Raabe, M., Baudry, N., Grillot, J-P. & Provansal, A. (1971) "Les organes périsympathiques des Insects Ptérygotes. Distribution. Caractères généraux," *C.r. hebd. Séanc. Acad. Sci., Paris* **273,** 2324–2327.
17. Highnam, K. C. & Goldsworthy, G. J. (1972) "Regenerated corpora cardiaca and hyperglycaemic factor in *Locusta migratoria," Gen. comp. Endocr.* **18,** 83–88.
18. Fraser Rowell, H. (1976) "The cells of the insect neurosecretory system: constancy, variability, and the concept of the unique identifiable neuron," *Adv. Insect Physiol.* **12,** 63–123.
19. Abbott, N. J., Moreton, R. B. & Pichon, Y. (1975), "Electrophysiological analysis of potassium and sodium movements in crustacean nervous system," *J. exp. Biol.* **63,** 85–115.
20. Evans, P. D., Kravitz, E. A. & Talamo, B. R. (1976), "Octopamine release at two points along lobster nerve trunks," *J. Physiol.* **262,** 71–89.

21. Bonga, S. E. W. (1971), "Formation, storage, and release of neurosecretory material studied by quantitative electron microscopy in the fresh water snail *Lymnaea stagnalis* (L.)," *Z. Zellforsch. mikrosk. Anat.* **113,** 490–517.
22. Joosse, J. (1976) "Endocrinology of molluscs." In *Actualités sur les hormones d'Invertébrés.* Colloques internationaux C.N.R.S.
23. Kanatani, N. (1976) "Hormones in Echinodermata." In *Actualités sur les hormones d'Invertébrés.* Colloques internationaux C.N.R.S.

Neurohypophysis

24. Barrington, E. J. W. (1975), *An Introduction to General and Comparative Endocrinology,* 2nd edition, Oxford, Clarendon Press.
25. Bandaranayake, R. C. "Morphology of the accessory neurosecretory nuclei and of the retrochiasmatic part of the supraoptic nucleus of the rat," *Anatomical Record* **80,** 14–22.
26. Nordmann, J. J. (1977) "Ultrastructural morphometry of the rat neurohypophysis", *J. Anat.* **123,** 213–218.
27. Olsson, R. (1959) "The neurosecretory hypothalamus system and the adenohypophysis of Myxine", *Z. Zellforsch. mikrosk. Anat.* **51,** 97–107.

Pineal organ

28. Collin, J. P., "The pineal organ of vertebrates", *Int. Rev. Cytol.* In press.
29. Ralph, C. L. (1975) "The pineal complex: a retrospective view," *Amer. Zool.* **15,** (suppl.), 105–116.

Chromaffin cells

30. Coupland, R. E. (1965). *The Natural History of the Chromaffin Cell,* London, Longmans.

Urophysis

31. Bern, H. A. (1969) "Urophysis and caudal neurosecretory system." In *Fish Physiology,* Vol. 2 (eds. W. S. Hoar & D. J. Randall), pp. 399–418, New York, Academic Press.

APUD system

32. Pearse, A. G. E. (1976) "Peptides in brain and intestine," *Nature* **262,** 92–94.

General reference for background reading

33. Gorbman, A. & Bern, H. A. (1962). *A Textbook of Comparative Endocrinology.* New York. John Wiley.

CHAPTER SIX

FUNCTIONS AND STRUCTURES OF NEUROHORMONES

IN CHAPTER FIVE WE WERE CONCERNED WITH THE WAYS IN WHICH THE neurosecretory systems of a variety of animals are organized in order to release neurohormones into the extracellular fluid. In this chapter we shall look in more detail at the hormones themselves to see what processes they control, how their secretion may be varied to produce different effects, and to find out what kinds of molecules neurohormones are. We shall consider first the situation in the invertebrates before looking at the neurohormones of vertebrates.

Processes and systems in invertebrates that are affected by neurohormones

This is not the place for an exhaustive consideration of all the processes known to be mediated by neurohormones. What may be of more general interest is to find out if there are definite kinds of processes which can be controlled more effectively by neurohormones than by, say, direct innervation. *A priori* it would seem more likely that processes under the control of neurohormones would be those in which the speed of response was not of overriding importance, and in which the response needs to persist for some considerable length of time. Obviously, the release of a neurohormone takes time to build up an effective concentration in circulation but, having reached such a useful concentration, this can be relatively easily maintained over extended periods. Within certain limits, these expectations are realized in what one finds.

Tables 6.1 and 6.2 list those processes for which there is evidence of their being under the control of neurohormones. It is striking that in non-arthropod invertebrates, virtually all the neurohormonal effects known are on morphogenetic processes; growth, development, gonad maturation and regeneration. Only in the molluscs and echinoderms is there evidence of control of such functions as egg laying, osmoregulation, and heart rate. To some extent, no doubt, this reflects the relatively undeveloped state of this area of investigation. After all, twenty-five years ago, virtually the only

Table 6.1 Processes for which there is evidence of control by neurohormones in non-arthropod invertebrates

Animal group	*Function*
Coelenterates	Promotion of growth and possible inhibition of sexual development Stimulation of head and bud formation
Platyhelminths	
—turbellarians	Control of regeneration Control of sexual development Promotion of asexual reproduction by division
Nematodes	Control of moulting fluid activation—possibly by causing an increased rate of water entry
Nemertines	Control of maturation of gonads
Echinoderms	Control of shedding of gametes
Annelids	
—polychaetes	Control of regeneration by proliferation of new segments Control of oocyte development and vitellogenesis Control of the metamorphosis from a sessile to a motile form
—oligochaetes	Control of regeneration Control of reproduction
—hirudinea	Control of gonadal development
Molluscs	
—gastropods	Control of spermatogenesis Control of growth Control of body water Control of ovulation and oviposition
—cephalopods	Control of heart beat

insect hormones known were those concerned with growth, moulting, and metamorphosis; the list now covers an enormously larger range of functions (Table 6.2). It is possible, though, that there is more to it than this. It seems possible, for example, that in animals with no or only a sluggish circulation of the body fluid, it would not be feasible to control with hormones processes which need regulation on a minute-to-minute basis. It is certainly the case that in insects, crustacea, and cephalopod molluscs, all of which have well-developed circulatory systems, many neurohormones are known which do have rapid effects and which are in circulation only for short periods of time.

Differential effects produced by modulation of hormone secretion

Two important effects of which Tables 6.1 and 6.2 give no hint are worth discussion. They concern the ways in which secretion of a single hormone can be varied so as to produce different effects.

Table 6.2 Processes in arthropods for which there is evidence of control by neurohormones

Group	*Function*
Crustaceans	Control of pigment granules in chromatophores
	Control of retinal pigment
	Control of succinate oxidation in mitochondria
	Control of water and ion levels
	Control of blood sugar levels
	Control of heart rate
	Control of gonadal development
	Control of moulting hormone release from Y-gland
	Control of glycogen synthetase and muscle phosphorylase
Insects	Control of ecdysone release from thoracic glands
	Control of breakdown of thoracic gland in adult
	Control of polymorphism of offspring
	Control of diapause in eggs
	Control of tanning of cuticle after ecdysis
	Control of breakdown of epidermal cells of wings
	Control of eclosion behaviour
	Control of ovulation
	Control of oviposition
	Control of colouring of pupa
	Control of colour change of epidermis
	Control of concentration of proteolytic enzymes in gut
	Control of protein synthesis in fat body
	Control of sugar concentration in haemolymph
	Control of lipid concentration in haemolymph
	Control of fluid secretion by Malpighian tubules
	Control of fluid absorption by rectum
	Control of absorption of fluid by midgut
	Control of fluid secretion by salivary glands
	Control of heartbeat
	Control of muscles of Malpighian tubules
	Control of visceral muscles

Since this table is intended only to give an indication of the types of process in which neurohormones are involved, some cases have been included for which the evidence of control by a neurohormone is not yet complete.

The first of these concerns the effect of changes in the *concentration* of hormone in circulation. A good example of this is the hormonal control of oocyte development in nereids (annelid worms), where a decreasing concentration of an inhibitory hormone controls all the phases of oocyte development.[1] At first, a high concentration of the hormone inhibits oocyte development. At the appropriate time its concentration declines, the oocytes begin to grow, and vitellogenesis is stimulated in an orderly fashion. The importance of the progressive and controlled decrease in the hormone levels is emphasized by the results of experiments with headless worms (the inhibitory hormone is released from the brain). Under these circumstances, the oocytes grow rapidly, but vitellogenesis is abnormal.

A rather similar example involves the control of metamorphosis in these nereid worms. At the time of sexual maturity, many nereid species change in form in rather a dramatic way. The parapodia and eyes enlarge greatly, and the animal becomes a much more actively swimming form. The process is known as *epitoky*. This change seems to be controlled by a steady decline in the level of hormone, which is inhibitory at high concentration, but which stimulates metamorphosis at lower concentrations.[2]

A second way in which a single hormone can be used to produce different effects is by variations in the *length of time* during which it is released. Probably the best example here is the different effects that follow different periods of release of the prothoracicotrophic hormone (PTTH) in the tobacco hornworm *Manduca sexta*.[3] PTTH controls the secretion of α-ecdysone (the precursor of the moulting hormone) by the prothoracic glands. In *Manduca*, as in many other Lepidoptera, the behaviour of the larvae changes dramatically towards the end of the last instar. They stop feeding to find and prepare a suitable site for pupation; only later on do they actually undergo pupation. It has now been discovered that this pattern of behaviour is caused by the secretion of PTTH for a first brief period of 3·5 hours, followed 2 days later by a longer period of PTTH secretion (*c.* 10 h), which causes the larva to pupate (figure 6.1). The first short burst of PTTH release is thought to induce a similarly brief period of

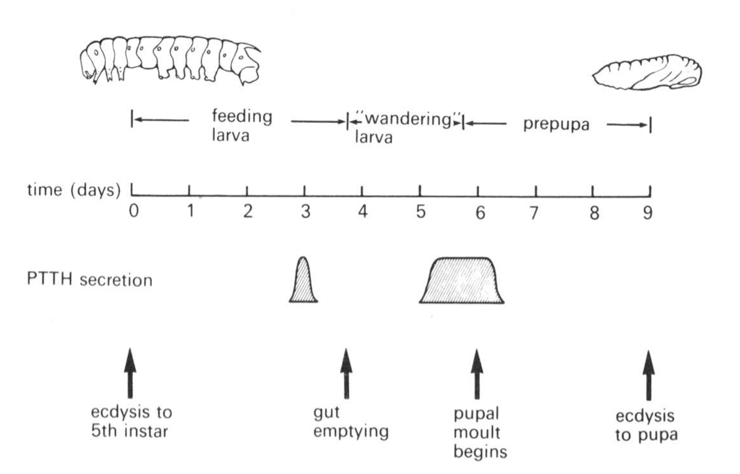

Figure 6.1 Events occurring during the 5th instar of *Manduca sexta*. Note the two periods of PTTH release of different length (after Nijhout and Williams, 1974, ref. 3).

ecdysone secretion by the prothoracic glands. This surge of ecdysone provokes the behavioural change away from feeding towards a phase characterized by an emptying of the gut, and by "wandering" behaviour as the larva searches for a pupation site. In addition, the larvae develop a pink pigment along the dorsal midline, and the tissues around the heart become transparent, so that it can be seen through the overlying cuticle. All these changes are thought to be caused by the short-lived exposure to ecdysone.

The second release of PTTH is more prolonged, and is thought to cause a longer and much larger release of ecdysone. In response to this, the larva moults to the pupal stage in the usual way.

This interesting use of two periods of PTTH secretion is just part of a fascinating series of regulated and coordinated changes, which it is worth pausing to describe, as it is an excellent example of control mechanisms at work.

When a feeding fifth-stage larva of *Manduca* reaches a weight of 5 g, endocrine glands known as the corpora allata stop secreting juvenile hormone (JH).[4] As a result, the circulating level of JH starts to fall and is undetectable by the time the larva weighs 7·5 g, which, since the larvae feed at a great rate, means a period of only just over 24 h. The presence of JH inhibits the brain from releasing PTTH from the corpora cardiaca (p. 90). In its absence, the brain becomes competent to secrete PTTH, which it

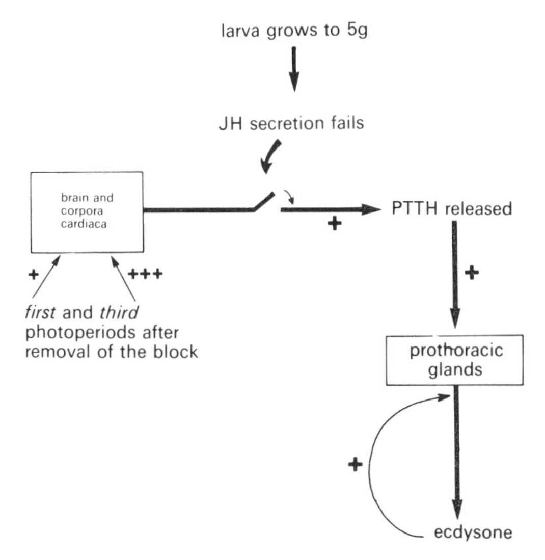

Figure 6.2 Hormonal events leading up to the pupal moult in *Manduca sexta*. + indicates a stimulatory effect.

is stimulated to do during the next photoperiod. As a result, a small amount of ecdysone is released by the prothoracic glands. Two days later, again in the light part of the day, PTTH secretion occurs again, this time for a longer period. This causes a release of ecdysone which is long enough to have a *positive feedback* effect on the prothoracic glands, so that the amount of ecdysone that is in circulation is considerably more than after the first PTTH release. This higher concentration leads to moulting rather than the changes observed 2 days earlier.

This whole chain of events, depicted in figure 6.2, illustrates many of the features that contribute to the flexibility and usefulness of a hormonal control system: inhibition, differential effects produced by changes in duration of release, differential effects produced by changes in concentration, as well as a feedback effect of the sort we shall discuss in more detail in chapter 8.

Multiple effects produced by single neurohomones

Neurohormones, as we shall see (p. 132), are relatively complex molecules capable of considerable specificity in the tissues that they affect. One might suppose, as a result, that a single neurohormone would only affect a single tissue. On the other hand we have seen (p. 127) how ecdysone (admittedly a "simpler" molecule than most neurohormones) has a variety of effects on a wide range of tissues.

In fact, it is now becoming clear that several neurohormones can each also affect more than one target tissue or organ. Most of the examples known concern insect neurohormones, but it is likely that other invertebrate neurohormones will be found to have such multiple effects. Teleologically speaking, the advantages that derive from such an arrangement are that, in this way, the different organs involved in a single physiological response can be coordinated in a "foolproof" way—there is no danger of any of the responding organs not being stimulated.

To see how this works let us look at two examples in the locust. The first concerns the adipokinetic hormone (AKH). For some time it has been known that AKH stimulates the release of lipid from the fat body during flight, so that it is available as an energy source. More recently, it has been discovered that the same hormone also stimulates the ability of the flight muscles to utilize the lipid which appears in circulation.[5] Clearly both processes are needed, and they can both be stimulated most economically by one hormone. A second example concerns the diuretic hormone of the locust. This promotes not only the secretory activities of the Malpighian

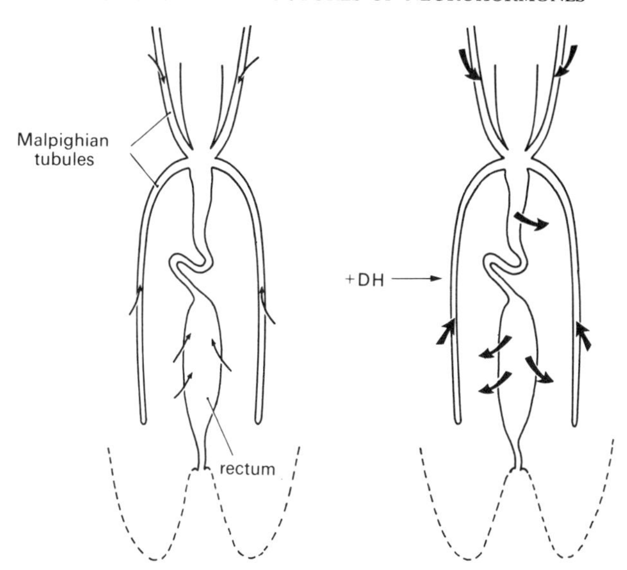

Figure 6.3 Concomitant increases in the rates of fluid secretion by the Malpighian tubules and fluid reabsorption by the hindgut of the locust in response to the appearance of the diuretic hormone (DH) in the haemolymph.

tubules, but also reabsorption in the hindgut. It is thought that during flight this hormone appears in circulation.[6] As a result, the haemolymph is more rapidly "filtered" through the Malpighian tubules, and its useful components (water, ions, amino acids and sugars) are recovered at appropriately higher rates by the hindgut (figure 6.3). In this way, useless or noxious products of the elevated metabolic activity occurring during flight are more rapidly removed from circulation, and yet the insect does not lose the valuable elements of its extracellular fluid.

A slightly more complex system occurs in the bloodsucking insect *Rhodnius,* and it illustrates not only the benefits of controlling several systems with one hormone, but also how differences in the dose/response parameter for each responding organ can provide additional advantages.

Rhodnius is a bloodsucking insect that takes very large meals—up to twelve times its own weight. It then excretes much of the excess water and ions (chiefly sodium and chloride) from the meal. This involves three epithelia (at least); the midgut wall absorbs a solution of NaCl into the haemolymph, the upper Malpighian tubules secrete an iso-osmotic fluid containing NaCl and KCl into the tubule lumen, and the lower Malpighian tubules reabsorb a hyper-osmotic KCl-rich fluid into the haemolymph.[7] As a result, a hypo-osmotic solution of NaCl is eliminated. For a short period after feeding, all these processes go on at rates much higher than before. All the evidence now suggests that the three epithelia are all controlled by one neurohormone, the *diuretic hormone*. The volume of the haemolymph and its potassium content are both closely controlled during diuresis. Yet either of these would be quickly affected by any imbalance between the activity of the midgut and of the different parts of the Malpighian tubules. For example, if the midgut absorbed fluid more slowly from the meal than the Malpighian tubules removed it, the haemolymph volume would rapidly decrease. That this does not happen seems to be attributable to a beautifully elegant arrangement of the details of the size and sensitivity of the response of these epithelia to the circulating hormone.

Figure 6.4 shows how the response of the midgut and the fluid-secreting part of the tubule varies with changes in the concentration of the hormone. The essential features are that the midgut can absorb fluid more rapidly than can the Malpighian tubules remove it, but that the Malpighian tubules are more sensitive to the hormone. What this means is that the haemolymph volume will change until the hormone concentration in it is such that the midgut and tubules transport fluid at the same rate. Given a constant amount of hormone in circulation, this situation is then stable. An increase in haemolymph volume, for example, will lower the hormone concentration, and the midgut will now absorb fluid more slowly than the tubules eliminate it, thereby reducing the haemolymph volume. A decrease in haemolymph volume causes an increase in hormone concentration, so that the midgut now absorbs fluid faster than the Malpighian tubules eliminate it, which restores the situation.

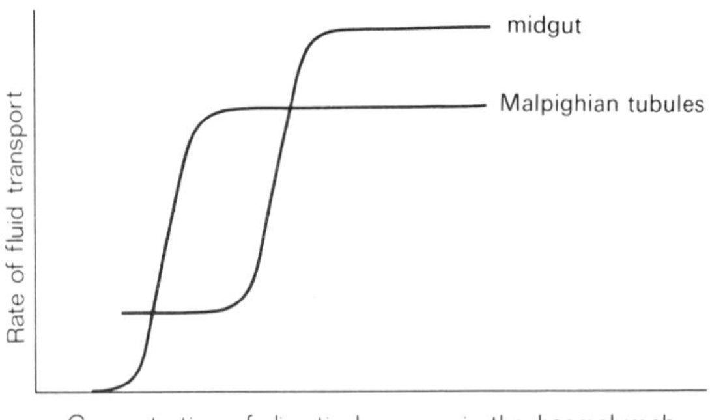

Figure 6.4 Dose/response curves for fluid transport by the midgut and Malpighian tubules of *Rhodnius* in the presence of diuretic hormone.

The potassium content of the haemolymph is kept close to 3–4 mM in spite of the fact that the upper Malpighian tubule secretes fluid containing 70 mM K. The lower tubule rapidly reabsorbs potassium (chloride) from the lumen. Both parts of the tubule are controlled by the same hormone. Plainly, the upper tubule ought not to be stimulated when the lower part is not, as this would lead to a very rapid loss of potassium from the haemolymph. In fact, the dose/response curves of the two regions of the Malpighian tubule are not identical, the lower tubule being more sensitive to the hormone. This means that the reabsorptive mechanism for potassium is always activated whenever fluid secretion by the upper tubule is stimulated. Although the same result could be achieved with two hormones, the use of only one eliminates the need for the much more complex control which would otherwise be needed.

The apparent disadvantage of controlling several different organs with one hormone is that the response would be a rather inflexible one. To take the example which we have just been discussing, it would seem that an increase in haemolymph potassium level in *Rhodnius* could not be eliminated by decreasing the level of hormone in circulation. While this would indeed cut down potassium reabsorption by the lower tubule, it would slow potassium elimination by the upper tubule to an even greater extent. The problem is solved in this case by a direct sensitivity of the lower tubule to the potassium level in the haemolymph. As the potassium level rises, the rate of potassium reabsorption slows, so that more is eliminated. This, together with the fact that a higher potassium level in the haemolymph leads directly to a more rapid secretion of this ion by the upper tubule, allows the concentration of potassium in the haemolymph to be regulated within narrow limits.

It is clear from all these examples that there are considerable advantages to be gained from a control system which involves the stimulation by a single hormone of a physiological response involving different tissues or organs.

The nature of invertebrate hormones

Although many processes in invertebrates are known to be controlled by neurohormones, not a great deal is known about the chemical nature of the hormones involved, and only very recently have the detailed structures of any of them been discovered. Table 6.3 sets out the information available on the chemical nature of a range of invertebrate hormones and Table 6.4 shows the detailed structure of the three hormones for which this

Table 6.3 The chemical nature of some invertebrate hormones

Animal	*Hormone*	*Nature of Hormone*	*Approx. Mol. Wt.*	*Threshold for Response*
Coelenterate				
Hydra attenuata	Head activator	peptide	1000	$<10^{-10}$M
Annelid				
Nereis diversicolor	controls oocyte development	peptide	≤ 1000	
Molluscs				
Aplysia californica	egg-laying hormone	peptide	6000	
Various cephalopods	cardioexcitor	peptide (destroyed by pronase)	1300	
Echinoderm				
Asterias amurensis	gamete-shedding hormone	peptide	2100	4×10^{-9}M
Crustaceans				
Cancer magister	hyperglycaemic hormone	peptide	6700	
Pandalus borealis	light-adapting distal retinal pigment hormone	peptide (18 amino acids)	2000	
Pandalus borealis	erythrophore-concentrating hormone	peptide (8 amino acids)	1000	
Crangon crangon	leucophore-concentrating hormone	peptide	1500	
Cancer borealis	cardioexcitor	peptide	1000	
Insects				
Periplaneta americana	cardioexcitor	peptide	1500-2000	
Leucophaea maderae	activator of hindgut muscle	basic peptide	500	
Bombyx mori	diapause hormone	contains peptide linkages	2000-4000	
Rhodnius prolixus	diuretic hormone	peptide	<2000	
Locusta migratoria	adipokinetic hormone	peptide	1158	5×10^{-9}M

Table 6.4 The structure of three arthropod neurohormones

(*a*) Light-adapting distal retinal pigment hormone of *Pandalus borealis* (Crustacea)	1 2 3 4 5 6 7 8 9 10 11 Asn-Ser-Gly-Met-Ile-Asn-Ser-Ile-Leu-Gly-Ile- 12 13 14 15 16 17 18 Pro-Arg-Val-Met-Thr-Glu-Ala-NH_2
(*b*) Erythrophore-concentrating hormone of *Pandalus borealis* (Crustacea)	1 2 3 4 5 6 7 8 pyroGlu-Leu-Asn-Phe-Ser-Pro-Gly-Tryp.-NH_2
(*c*) Adipokinetic hormone of locust (Insecta)	1 2 3 4 5 6 7 8 9 10 pyroGlu-Leu-Asn-Phe-Thr-Pro-Asn-Tryp-Gly-Thr-NH_2

Note the strong similarities in structure between hormone (b) & (c)—see text.

information is so far available. Some points are worthy of comment. The first is that nearly all invertebrate neurohormones are protein or peptide in nature, with molecular weights usually in excess of 1000. Peptides of this size presumably are large enough to be highly specific in their effects, so that receptor sites are able to distinguish different hormones.

The second point emerges from comparing the structures of the adipokinetic hormone (AKH) of the locust with that of the erythrophore-concentrating hormone of the crustacean *Pandalus borealis.* As Table 6.4 shows, the amino acids occupying positions 1–4, 6 and 8 are the same in the two hormones. As a result, as one might expect, these two hormones have cross effects when injected into the "wrong" organism.[8] Locust AKH, for example, causes blanching when injected into the crustacean, even at a dose as low as 1 pmol per animal. The similarity in the structure of these two hormones has to be set against the fact that in Crustacea the hormone is used as an agent to control colour change, while in insects the other hormone is used in controlling lipid movements. The situation is about what one would expect in two groups which, although they are phylogenetically related, have an evolutionary divergence of long standing. They do, however, have in common β-ecdysone as the moulting hormone; no differences in structure have accumulated in this case—perhaps because of the small molecular size of ecdysone which might not allow it to accommodate structural changes without losing much of its activity.

Finally, Table 6.3 shows that, on the limited evidence available, invertebrate hormones seem not to be as active as those of vertebrates. We saw earlier (p. 24) that the mammalian hormone oxytocin is active in the range 10^{-12}—10^{-13}M, whereas lowest AKH is active at concentrations around 10^{-8}M.[8] It is easy to see that this might be correlated with the great difference in volume of the extracellular fluid in the two animals, with the larger one needing either more hormone or, more economically, a more active hormone in circulation to have its effect.

Vertebrate neurohormones

Tables 6.1 and 6.2 showed the impressively large numbers of processes controlled by neurohormones in invertebrates. The situation in vertebrates is dramatically different. Apart from the hypophysiotrophic neurohormones which have essentially local actions, only three vertebrate neurohormones have been shown by good evidence to appear in general circulation. These are oxytocin, vasopressin, and adrenaline which are released from the neurohypophysis (oxytocin and vasopressin) and from the chromaffin cells (adrenaline). There is some evidence, not yet complete, that neurohormones are released from the pineal gland and from the urophysis of lower vertebrates; we shall give this evidence and attempt to evaluate it later in the chapter (p. 137).

Hypothalamic regulatory peptides

Although only a few neurohormones appear at effective concentrations in the general circulation of vertebrates, there are a growing number of neurosecretory factors known to be released locally into the hypothalamic-hypophysial blood portal system.[9] These factors act as local extracellular signals or messengers. They act to regulate the release of hormones from the anterior pituitary (adenohypophysis), which of course are not neurohormones. It is not easy at the moment to suggest why vertebrates should have evolved such a relay system of local neurosecretory control of non-neurosecretory hormone-producing tissues, and that invertebrates tend on the contrary to use neurohormones directly to control so much of their physiology. Suggestions, which may be worth consideration, include the possibilities that a control process involving more than one step may be more flexible, more open to feedback control, and may more easily allow amplification than a more straightforward solution. For large terrestrial vertebrates, in particular, amplification of a limited signal from the nervous system would allow the release of sufficiently large amounts of hormone without correspondingly large breeches in the vital blood-brain barrier.

These locally-released substances are known as *hypothalamic regulatory peptides* or *hypophysiotrophic hormones* or *releasing factors*. They are synthesized by neurones in the hypothalamus known as the *parvocellular*

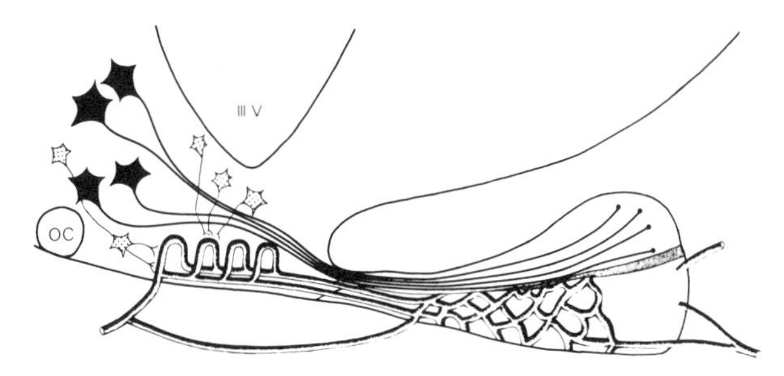

Figure 6.5 To show the course of the axons of the neurosecretory neurones of the hypothalamus in a mammal. The axons of the oxytocin and vasopressin-secreting cells (cell bodies in black) run down and end in the neurohypophysis. The axons of the neurones of the parvocellular system (cell bodies drawn cross-hatched) end on the primary capillary plexus of the hypothalamic-hypophysial portal system (at left).

system. Their nerve endings are to be found in the median eminence of the brain. The neurosecretory products are released into the interstitial spaces adjacent to the primary capillary plexus of the hypothalamic-hypophysial portal system (figure 6.5). They thus have virtually immediate access to their target cells in the anterior pituitary. The neurones releasing these regulatory factors are thought in turn to be under the control of specialized neurones secreting a range of neurotransmitters including dopamine, noradrenaline, serotonin (5-HT) and acetylcholine.[10] Table 6.5 lists the regulatory peptides involved in the system and shows the detailed amino-acid sequence in the three cases where this has been established.

Because their detailed structure is now known, it has been possible to synthesize these peptides and investigate their effectiveness at various concentrations. It turns out that the threshold concentration for any effect is in the range of 30–100 pM and half-maximal effects are achieved[10] at 0·5–2 nM. These values are rather higher than those required for, say, oxytocin and it may well be that, as before, the highest sensitivities are not needed in a system where the neurosecretory material is released into a relatively small volume.

Thyrotropin releasing hormone (TRH) is a relatively simple compound of low molecular weight (Table 6.5) and this may explain the strange fact that it has even been found in the circumoesophageal ganglia of the snail;[11] obviously, the simpler the peptide, the more chance there is that it may have been evolved more than once for use as an extracellular messenger.

One oddity which has emerged recently is that several of these supposedly hypothalamic peptides have been found in other parts of the

Table 6.5 Hypothalamic regulatory peptides and their structures

Factor	*Effect*	*Structure*
Growth hormone releasing factor (GRF)	Causes release of growth hormone	
Thyrotropin-releasing hormone (TRH)	Causes release of thyrotropin (TSH)	1 2 3 pyroGlu-His-Pro-NH_2
Somatostatin (SS)	Inhibits TRH-induced TSH release	1 2 3 4 5 6 7 8 H-Ala-Gly-Cys-Lys-Asn-Phe-Phe-Trp (Cys 3 linked to Cys 14; Trp 8 linked to Lys 9) HO-Cys-Ser-Thr-Phe-Thr-Lys 14 13 12 11 10 9
Luteinizing hormone releasing hormone (LRH)	Causes release of luteinizing hormone (LH)	1 2 3 4 5 6 7 8 9 10 pyroGlu-His-Trp-Ser-Tyr-Gly-Leu-Arg-Pro-Gly-NH_2
* Corticotropin-releasing factor	Causes release of corticotropin	
*Melanocyte-stimulating hormone inhibitory factor (MIH)	Inhibit release of melanocyte stimulating hormone (MSH)	

*The status of these two factors is still in doubt.

body—notably in cells derived from neural crest material and now lying in the gut and pancreatic islets, i.e. cells forming part of the APUD series[12] (p. 120).

Possible neurohormones from the pineal gland and urophysis

Oxytocin and vasopressin and their structural relatives found in other vertebrates have well-established roles as neurohormones. Before we consider the processes that these hormones control, what of the neurosecretory products of the pineal gland and urophysis (see pp. 104 and 116)?

As we saw earlier (p. 108) there is good evidence that the *pineal gland* synthesizes melatonin. In the so-called lower vertebrates this compound is probably involved in the control of pigmentation. In mammals, melatonin synthesis is a response to noradrenaline, a neurotransmitter released from postganglionic sympathetic nerves. In man, the level of noradrenaline declines in the light, so that most melatonin is secreted during sleep.[13] In mammals, melatonin has antigonadal properties, probably due to its action in reducing the release of luteinizing hormone (LH). Recent work has shown that the pineal gland also synthesizes peptides, among them being arginine vasotocin (AVT-see Table 6.6). Interestingly it has been found that AVT release can be stimulated by melatonin.[14] AVT has similar actions to melatonin but is about one million times more potent.[14] It is possible therefore that melatonin may not, in mammals, be the effective agent of the pineal gland but only an intermediate compound in a chain of events culminating in AVT release.

Most of the evidence that the *urophysis* is an active endocrine source is based on its structure, which is clearly similar to other neurohaemal organs. However, several different physiologically active peptides can be isolated from urophyses; the evidence that these factors are used as hormones has received support from recent studies. At least one of the active peptides is present at high concentration in a fraction prepared from trout urophyses by differential centrifugation and known to be rich in neurosecretory granules. At least two active materials are released from isolated intact urophyses during treatment with solutions containing elevated concentrations of potassium, and the release is calcium-dependent. As we have seen (p. 47), such behaviour is characteristic of neurohaemal organs. Finally, the amount of extractable peptides in the urophysis is many times higher than that needed to have pronounced physiological effects, so they could well be used as hormones. What functions the urophysial hormones

Table 6.6 The structure of the various peptides found to occur in vertebrate neurohypophyses

Hormone	1 Cys	-	2 Tyr	-	3 ⓐ	-	4 ⓑ	-	5 Asn	-	6 Cys	-	7 Pro	-	8 ⓒ	-	9 Gly-NH_2
Oxytocin (OT)					Ile		Gln								Leu		
Arginine vasopressin (AVP)					Phe		Gln								Arg		
Arginine vasotocin (AVT)					Ile		Gln								Arg		
Lysine vasopressin (LVP)					Phe		Gln								Lys		
Mesotocin (MT)					Ile		Gln								Ile		
Isotocin (IT)					Ile		Ser								Ile		
Glumitocin (GT)					Ile		Ser								Gln		
Aspartocin (AT)					Ile		Asn								Leu		
Valitocin (VT)					Ile		Gln								Val		

Table 6.7 The occurrence of neurohypophysial peptides in the different classes of vertebrates (from Pickering, 1978, Ref. 20)

	Fish											
Peptide	Cyclostomes	Chimaeroids	Sharks	Rays	Lungfish	Polypterus	Sturgeon	Teleosts	*Amphibians*	*Reptiles*	*Birds*	*Mammals*
Oxytocin (OT)		+			+				possibly	possibly	possibly	+
Arginine Vasotocin (AVT)	+	+	+	+	+	+	+	+	+	+	+	
Mesotocin (MT)					+	possibly			+	+	+	
Isotocin (IT)						+		+				
Glumitocin (GT)				+								
Aspartocin (AT)			+									
Valitocin (VT)			+									
Arginine Vasopressin (AVP)												+
Lysine Vasopressin (LVP)												+

might have is still controversial. Such evidence as there is suggests roles in osmoregulation, control of the cardiovascular system, and in affecting the activity of smooth muscle.[15]

The nature of hormones from the neurohypophysis of vertebrates

In chapters 2, 3, & 4 we dealt with the synthesis, axonal transport, and release of the mammalian neurohormones, oxytocin and vasopressin. In fact these hormones are not universal in vertebrates; the non-mammalian vertebrates have an array of different peptide neurohormones present in their neurohypophysial tissue. The first one to be investigated was a peptide, arginine vasotocin (AVT), found in the neurohypophyses of teleosts, amphibians, reptiles, and birds. Most interestingly, it turned out to have a structure halfway between that of oxytocin and vasopressin (see Table 6.6) with the ring of oxytocin but the side chain of vasopressin.[16] Not surprisingly, in mammalian assays, it has something of the properties of both the mammalian hormones but with lower potency. Vasopressin (ADH) is well known to affect water movement through the skin of frogs. Fittingly, AVT, which occurs naturally in amphibia, is more than a hundred times as potent in its effects on transport processes in the skin of frogs.[17]

Since the discovery of AVT, several related peptides have been found in vertebrate neurohypophyses. Their names, structure and distribution are shown in Tables 6.6 and 6.7. It is clear from these that naturally occurring variations are found only in positions 3, 4 and 8 of the basic structure. AVT is very widespread, but apart from mesotocin, the others are each found in only one or two groups. AVT has even recently been found in the neurohypophyses of foetal mammals.[18] Mesotocin is interesting in that it occurs in all the non-mammalian tetrapods, i.e. amphibians, reptiles and birds, and has been found in lungfish which are thought to be closely related to the ancestors of the amphibians. Lysine vasopressin (LVP) is found in the suborder Suina of the artiodactyls (this group includes pigs, peccaries, and the hippopotamus). Interestingly, some individuals of the pig family are homozygous for LVP, others are homozygous for arginine vasopressin (AVP), while still other are heterozygous and have both AVP and LVP. Such polymorphism casts some doubt on attempts to reconstruct the phylogeny of the vertebrate neurohypophysial hormones. Schemes of this nature are, however, of great interest and provided they are interpreted with care they can be illuminating. One such scheme is shown in figure 6.6. It proposes that AVT was the first evolved vertebrate neurohypophysial

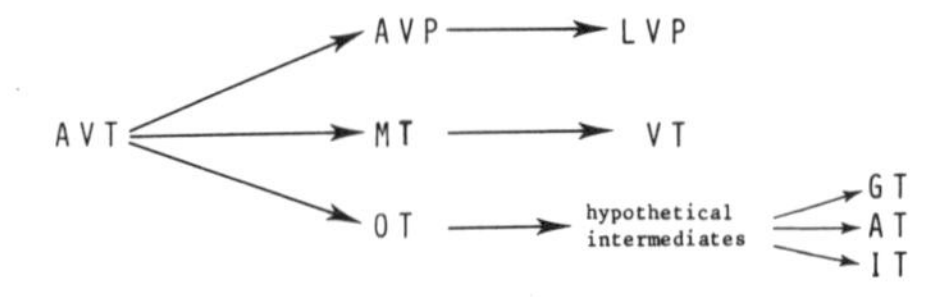

Figure 6.6 A hypothetical scheme for the evolution of vertebrate neurohypophysial peptides (after Heller & Pickering, 1970, ref. 16).

hormone. From this, by gene duplication, two different lines of peptides were evolved: a series of basic peptides which has changed little in evolution, and which comprises AVT and the mammalian vasopressins (LVP and AVP), and a series of oxytocin-like peptides which has evolved more rapidly.

So much for the structure and nature of the hormones released into circulation by the neurohypophysis. We must now consider their functions.

Physiology of the neurohypophysial hormones in mammals

Oxytocin

In most mammals, oxytocin plays a central role in lactation and may also be involved in parturition.[19] In lactation it produces milk ejection by causing contraction of the myo-epithelial cells which surround the alveoli and ducts of the mammary gland. Suckling produces milk ejection by a reflex acting through the central nervous system and leading to oxytocin release (figure 6.7). In parturition its role is not clear. Although uterine contractions can be induced by oxytocin injections, naturally occurring labour in women seems not to cause any increase in pressure in the ducts of the mammary gland, although this is very sensitive to oxytocin release. In other mammals, however, oxytocin has been detected in the blood during parturition.

Vasopressin

As is well known, the function of vasopressin is to play a part in the homeostatic control of the volume of extracellular fluid (figure 6.7).[19] It acts to conserve body water by reducing the flow of urine produced by the kidneys. This it does by increasing the osmotic permeability of the terminal part of the distal convoluted tubules and, more importantly, of the collecting ducts. As a result, the hypo-osmotic fluid flowing through the distal convoluted tubule rapidly loses water to the interstitial fluid and becomes iso-osmotic with it. During its subsequent passage through the collecting duct, the luminal fluid loses still more water as it equilibrates with the progressively higher osmotic concentration of the interstitial fluid in the medulla. The resulting urine is not only very concentrated, but its rate of production is very much slowed.

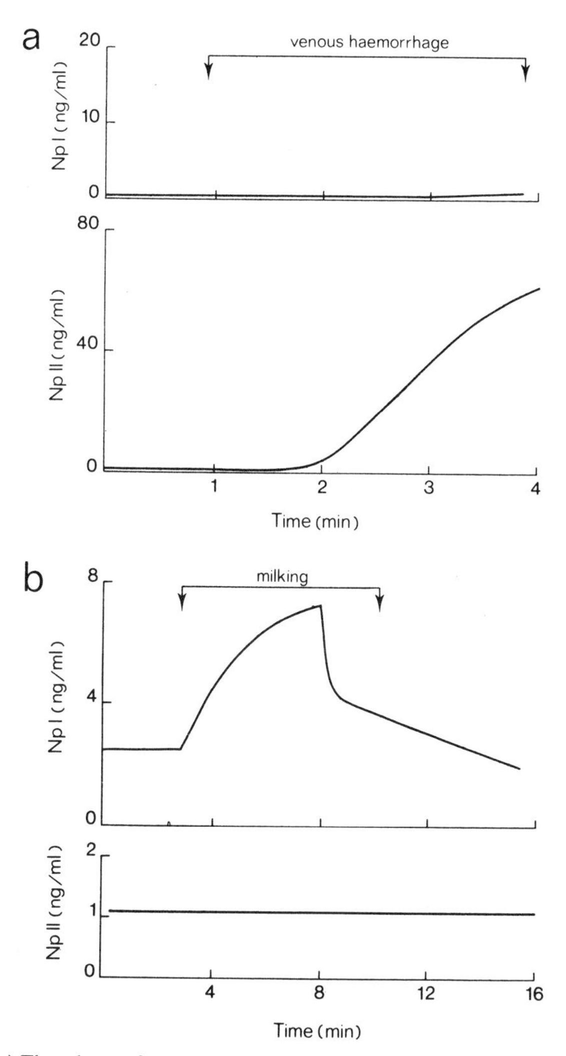

Figure 6.7 (*a*) The release of neurophysin II, NpII, (which is associated with AVP) during haemorrhage and (*b*) the release of neurophysin I, NpI, (which is associated with oxytocin) during milking in a mammal. Note that in both cases one neurophysin, i.e. one hormone only, is released (*a*, after Legros *et al.*, 1975, ref. 21; *b*, after Legros *et al.*, 1974, ref. 22).

The action of neurohypophysial hormones in non-mammalian vertebrates

Since most non-mammalian vertebrates have two neurohypophysial peptides, it is tempting to suggest that they might act as in mammals, i.e. one with a primarily antidiuretic effect and one acting in the control of reproduction. Such a suggestion is not supported by the facts, however, as it seems that arginine vasotocin (AVT) has both sorts of effects in most species, while the role of the other peptide remains obscure.

AVT is considered to be the non-mammalian antidiuretic hormone, e.g. it increases water absorption in the kidneys of amphibians, reptiles, and birds.[20] In addition, it depresses urine flow in amphibians and reptiles by decreasing the glomerular filtration rate. So, just as we saw for some invertebrate hormones (p. 128), the hormone appears to have more than one target, though both are concerned in the one physiological response. In fact, the effects of AVT in antidiuresis may go wider than this. In amphibians, in addition to the kidney, the bladder wall and the skin are important sites of absorption of sodium and, in the bladder, of water as well. AVT has potent effects on transport processes in these tissues.[20]

In fish, AVT has an interesting effect. At doses above 10^{-11}M, AVT has a diuretic effect, but it causes antidiuresis at levels in the range 10^{-12}—10^{-15}M.[20] Again this is reminiscent of the differential effects found with some invertebrate hormones which have different actions depending on their concentration (pp. 125, 130).

With one exception, none of the other neurohypophysial peptides occurring in non-mammals is as effective as AVT in the processes described. The exception is mesotocin (Table 6.6), which causes more net uptake of sodium across the gills of fish than does AVT. It does appear that AVT is the non-mammalian antidiuretic hormone acting to conserve body water.

AVT causes contraction of the oviducts in the chicken, and neurohypophysial extracts induce similar effects on the isolated oviducts of fish, amphibians, and reptiles. In addition, the amount of AVT in the neurohypophysis of the hen declines during oviposition, and the oviducts of fish and amphibians become most sensitive to AVT just before the eggs or young are expelled. These findings all suggest that AVT has a role in controlling reproduction in these animals.

Strangely then, AVT seems not only to regulate body water and electrolyte levels in non-mammals, but also to regulate the action of the reproductive tracts. How these actions are separately controlled is not clear. Also not clear is the role, if any, of the second peptide which occurs in so many non-mammalian neurohypophyses.

REFERENCES

1. Clark, R. B. (1965) "Endocrinology and reproductive biology of polychaetes." In *Oceanography & Marine Biology, An Annual Review* **3,** (Ed. H. Barnes), London, Allen and Unwin.
2. Durchon, M. (1976) "Les hormones chez les Vers et les Annélides." In *Actualités sur les hormones d'Invertébrés.* Colloques internationaux C.N.R.S.
3. Nijhout, H. F. & Williams, C. M. (1974) "Control of moulting and metamorphosis in the tobacco hornworm, *Manduca sexta* (L.): growth of the last-instar larva and the decision to pupate," *J. exp. Biol.* **61,** 481–491.
4. Nijhout, H. F. & Williams, C. M. (1974) "Control of moulting and metamorphosis in the tobacco hornworm, *Manduca sexta* (L.): cessation of juvenile hormone secretion as a trigger for pupation," *J. exp. Biol.* **61,** 493–501.
5. Robinson, N. L. & Goldsworthy, G. J. (1977) "Adipokinetic hormone and the regulation of carbohydrate and lipid metabolism in a working flight muscle preparation," *J. Insect Physiol.* **23,** 9–16.
6 Goldsworthy, G. J. (1976) "Hormones and flight in the locust." In *Perspectives in Experimental Biology,* Vol. 1, Zoology (Ed. P. Spencer Davies), Oxford, Pergamon Press.
7. Maddrell, S. H. P. & Phillips, J. E. (1975) "Secretion of hypo-osmotic fluid by the lower Malpighian tubules of *Rhodnius prolixus,*" *J. exp. Biol.* **62,** 671–683.
8. Mordue, W. & Stone, J. V. (1977) "Relative potencies of locust adipokinetic hormone and prawn red pigment-concentrating hormone in insect and crustacean systems," *Gen. Comp. Endocrinol.* **33,** 103–108.
9. Yates, F. E., Russell, S. M. & Maran, J. W. (1971) "Brain-adenohypophysial communication in mammals," *A. Rev. Physiol.* **33,** 393–444.
10. Reichlin, S., Saperstein, R., Jackson, I. M. D., Boyd, A. E. & Patel, Y. (1976) "Hypothalamic hormones," *A. Rev. Physiol.* **38,** 389–424.
11. Grimm-Jorgensen, Y., McKelvy, J. F. & Jackson, I. M. D. (1975) "Immunoreactive thyrotropin releasing factor in gastropod circumoesophageal ganglia," *Nature* **254,** 620–621.
12. Pearse, A. G. E. (1976) "Peptides in brain and intestine," *Nature Lond.* **262,** 92–94.
13. Anon.(1977) "Current knowledge on the pineal organ," *Research in Reproduction* **9,** (4), 3–4.
14. Pavel, S. (1973) "Arginine vasotocin release into cerebrospinal fluid of cats induced by melatonin," *Nature, New Biol.* **246,** 183–184.
15. Berlind, A. (1973) "Caudal neurosecretory system: a physiologist's view," *Amer. Zool.* **13,** 759–770.
16. Heller, H. & Pickering, B. T. (1970) "The distribution of vertebrate neurohypophysial hormones and its relation to possible pathways for their evolution." In *International Encyclopaedia of Pharmacology and Therapeutics,* Section 41, Volume 1, Oxford & New York, Pergamon.
17. Pickering, B. T. (1970) "Aspects of the relationships between the chemical structure and biological activity of the neurohypophysial hormones and their synthetic structural analogues." In *International Encyclopedia of Pharmacology and Therapeutics,* Section 41, Volume 1, Oxford & New York, Pergamon.
18. Perks, A. M. & Vizsolyi, E. (1973) "Studies on the neurohypophysis in foetal mammals." In *Foetal and Neonatal Physiology,* pp. 430–438, Cambridge University Press.
19. Bisset, G. W. (1976) "Neurohypophysial hormones." In *Peptide Hormones,* (ed. J. A. Parsons), London, Macmillan.
20. Pickering, B. T. (1978) "Posterior-lobe hormones—Comparative Aspects." In *Hypothalamic Hormones.* New York, Academic Press.
21. Legros, J. J., Reynaert, R. & Peeters, G. (1975) "Specific release of bovine neurophysin II during arterial or venous haemorrhage in the cow," *J. Endocrinol.* **67,** 297–302.
22. Legros, J. J., Reynaert, R. & Peeters, G. (1974) "Specific release of bovine neurophysin I during milking and suckling in the cow," *J. Endocrinol.* **60,** 327–332.

CHAPTER SEVEN

MECHANISMS OF NEUROHORMONE ACTION

The mechanism of hormonal action

So far we have seen how neurosecretory hormones are synthesized, stored and released, and the sorts of effects they give rise to. In doing this, we have missed out one step. How are hormones able to influence their target cells? Neurohormones are usually fairly large compounds which, one would expect, could not easily find their way into the cells which they affect. Indeed, what happens is that the hormone molecules "recognize" and attach themselves to receptor molecules on the cells' plasma membrane. Having alighted on the cell surface, they then induce changes in or close to the membrane and this, in turn, gives rise to the whole series of intracellular processes constituting the hormonal effect. Even where the hormone or chemical transmitter is as small a molecule as noradrenaline or acetylcholine, the same sort of mechanism operates, presumably in this case in the interests of speed; acetylcholine, for example, can bring about a change in permeability of the postsynaptic membrane within a period of less than one millisecond. On p. 149 we deal with the question of the membrane changes which, in essence, transfer information across the cell membrane. First, however, we should describe the interaction of hormones with their receptor sites.

At one time it was thought that neurohormones (and indeed other hormones such as insulin) became bound to their receptors by an essentially irreversible process such as the formation of a covalent bond or bonds. Reversal of stimulation would require that the hormone was later removed in some way by being broken down or by being absorbed into the cell, or in some other fashion. If this did not happen, more and more hormone would bind to receptors until they were saturated, and the response of the cell would quickly become maximal.

More recent research has shown that such strong binding does not occur. Hormone molecules in fact bind rather loosely to their receptors. They associate and dissociate with relative ease, and at rates which follow the

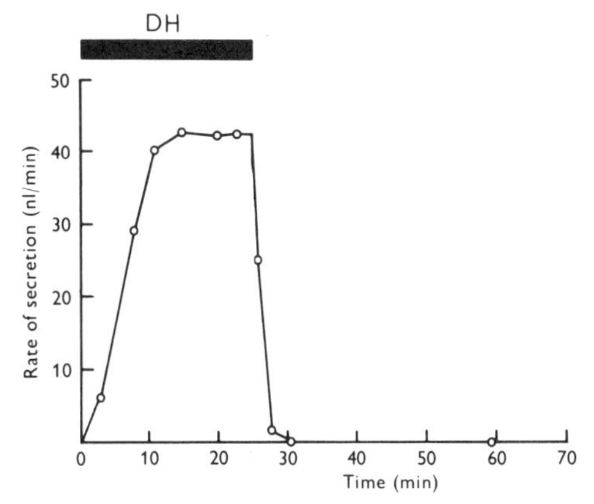

Figure 7.1 The rate of fluid secretion by an isolated Malpighian tubule of *Rhodnius* during and after exposure to the diuretic hormone (DH).

kinetic characteristics of simple dissociation processes. Those hormone molecules which are bound to receptors and that dissociate into the medium are immediately able to bind again to the same or other receptors and continue their stimulant activity.

Two examples from recent work on insect systems will serve as illustrations.[1,2] The diuretic hormone of the bug *Rhodnius* causes rapid fluid secretion by the insect's Malpighian tubules. When the hormone-containing medium is replaced by one containing no stimulant, the rate of fluid secretion rapidly falls (figure 7.1); the hormone dissociates from the receptors and is easily washed away. The tubules will secrete only slowly in a medium containing a stimulant if there is also present a compound such as tryptamine which acts as a competitive inhibitor, and they only slowly increase their rate of fluid secretion if they are returned to a medium containing only a stimulant (figure 7.2). In this case it appears that the inhibitor detaches from the receptor sites at such a low rate that the number of sites available to the hormone increases only slowly to its normal level. Surprisingly enough, a successful stimulant such as 5-HT is effective at 3×10^{-8}M, while 10^{-5}M tryptamine is needed to have an inhibitory effect. It seems that after the initial association of a stimulant molecule with its receptor there must be a sharp fall in the affinity between them, presumably due to a conformational change in the receptor. Competitive inhibitors have a lower affinity for unoccupied receptor sites, but once bound to them they have a strong affinity for the altered receptors, and so dissociate from them only with reluctance. The sequence of events that one envisages is along the lines shown in figure 7.3.

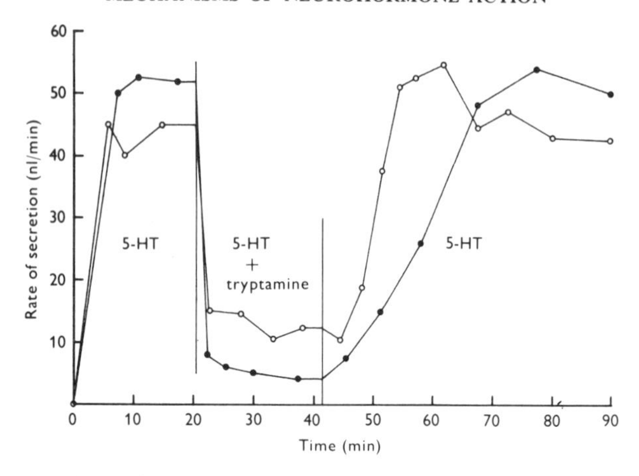

Figure 7.2 The effects of a period of immersion in 8×10^{-5}M tryptamine (open circles) or $6{\cdot}5 \times 10^{-4}$M tryptamine (filled circles) on the rates of secretion of Malpighian tubules of *Rhodnius*, induced to secrete in 10^{-5}M 5-HT.

If the cycle of events can proceed rapidly, then rapid secretion of fluid follows. Because inhibitory molecules leave the receptor sites reluctantly, the cycle is slowed down in their presence, and so secretion slows down or stops. Although tryptamine is an inhibitor, it does slowly leave the receptors and, if it is present at concentrations high enough to encourage its attachment to reactivated receptors (step 1 in figure 7.3), there is a slight stimulant effect on fluid secretion by the tubules. Presumably under these conditions, the cycle of receptor events can proceed slowly. 5-methyltryptamine is much less easily washed off the receptors than is

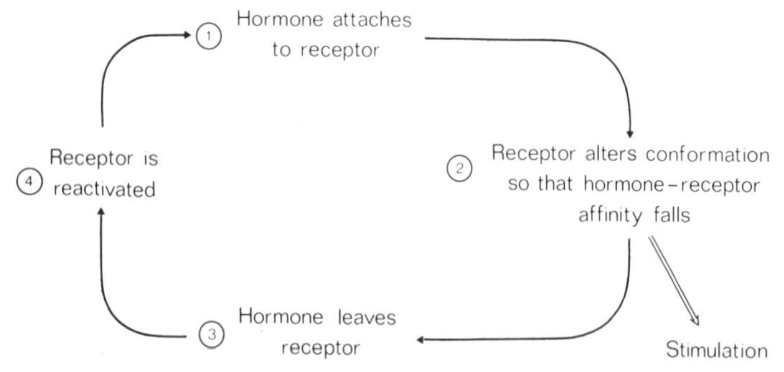

Figure 7.3 Events thought to underlie the stimulatory action of the diuretic hormone of *Rhodnius* (after Maddrell *et al.*, 1971, ref. 1).

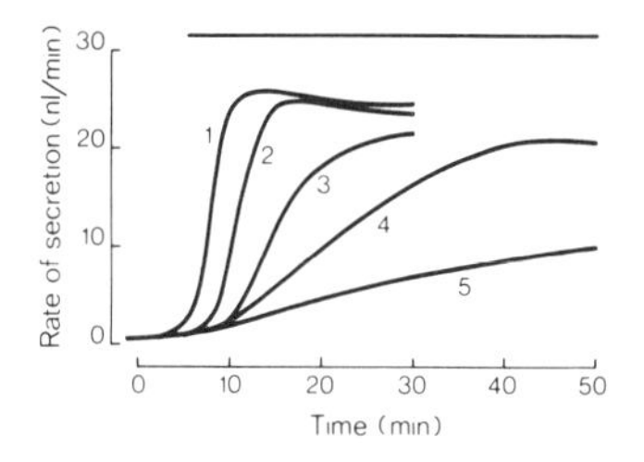

Figure 7.4 The effect of different concentrations of LSD on the speed of onset of fluid secretion by isolated salivary glands of *Calliphora erythrocephala*. LSD was present during the time indicated by the bar. The concentrations of LSD used were 1; 100nM; 2, 1nM; 3, 100 pM; 4, 10 pM; 5, 5pM (after Berridge & Prince, 1974, ref. 2).

tryptamine and so, as one might predict, very soon there is no fluid secretion at all in its presence. Even with 5-methyltryptamine, however, an initial treatment at 10^{-2}M causes a short-lived burst of secretion by the tubules. Presumably many inhibitor-receptor interactions occur almost simultaneously, and proceed as far as step 2 of figure 7.3, and this is sufficient for a short-lived stimulation. It seems most likely in fact that step 2 is the one which leads to stimulation, as a change in the configuration of the receptor can easily be imagined to have the sorts of consequence which lead to the hormonal effect.

Work on the effects of stimulants on fluid secretion by blowfly salivary glands (which is also stimulated by 5-HT) has added several interesting points. Perhaps the most illuminating case concerns the action of lysergic acid diethylamide (LSD). On treatment with very low concentrations of LSD (down to 5×10^{-12}M), the glands start to secrete fluid initially at a very low rate but the rate gradually increases (figure 7.4). The LSD is not easily washed off, so that secretion by LSD-treated glands returned to a stimulant-free medium only slowly declines to lower levels (figure 7.5). To

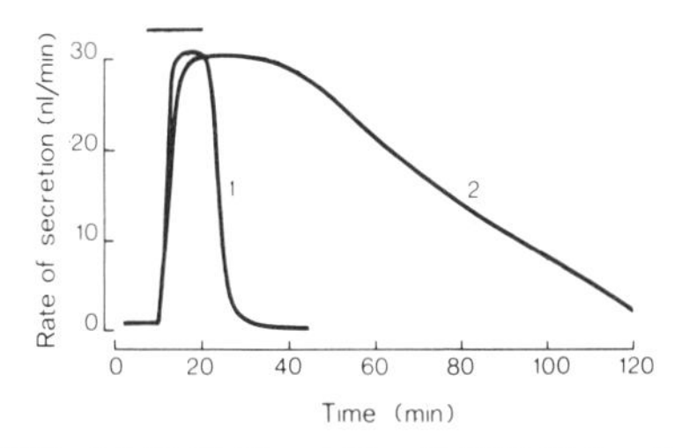

Figure 7.5 The effect of 10 mM 5-HT (1) and 10 mM LSD (2) (applied for the time indicated by the bar) on the rate of fluid secretion by isolated salivary glands of *Calliphora erythrocephala* (after Berridge and Prince, 1974, ref. 2).

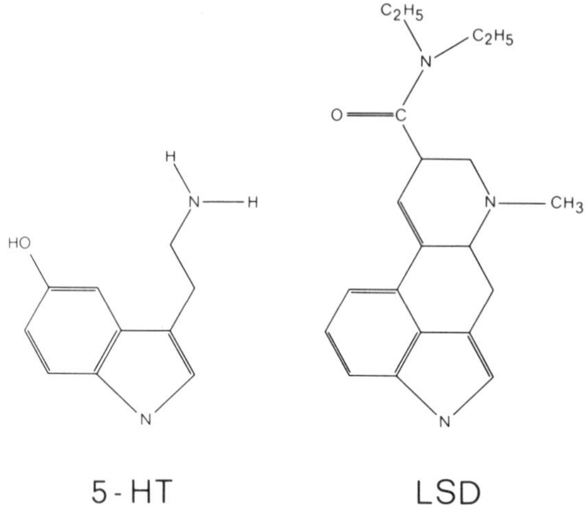

5-HT LSD

Figure 7.6 The structures of 5-HT and LSD.

accommodate this unusual behaviour in the scheme shown in figure 7.3 one has to suppose that LSD might strongly bind to additional sites close to the active part of the 5-HT receptor. This binding would allow the free parts of the LSD molecule repeatedly to attach to and detach from the 5-HT receptor in the manner suggested before. As shown in figure 7.6, part of the structure of LSD is similar to 5-HT but there are additional regions which might be involved in binding to other sites close to, but not actually part of, the 5-HT receptor. The stimulant action of very low concentrations of LSD would involve the following sequence of events. Initial attachment would occur via those parts of the molecule other than that resembling 5-HT. This binding is not easily reversed, so more and more LSD molecules would become bound in this way. The number of 5-HT-like regions which would now be in the right place to stimulate 5-HT receptors would therefore steadily increase, leading to progressively higher rates of fluid secretion.

Evidence in favour of such an interpretation is that an inhibitor such as gramine can still slow the secretion of LSD-stimulated glands. When the active part of the LSD molecule detaches from the receptor, there is a brief period when the receptor is available for other compounds to attach and although they have to compete with LSD molecules held poised over the receptor, they can still gain access to the receptor if they are present at a high enough concentration.

In that any changes to the diethylamide group of LSD drastically reduces its effects, it seems likely that it is this group which is at least partly responsible for the binding at sites adjacent to the 5-HT receptor.

In summary, then, neurohormones stimulate the cells of their target organs by transient binding to specific receptor sites at the cell surface. The short-lived nature of the association ensures that the response of the cells can be regulated by changes in the hormone concentration, and also that the system can rapidly be switched off when the hormone no longer appears in the surrounding medium.

Second messengers

We now turn to the question of just how hormone-receptor interactions lead to the cellular response. In many cases it is known that the response occurs in parts of the cell away from those areas accessible to the hormone. Some kind of link is necessary to overcome this. It is increasingly often found that after the reception of information at the cell surface, there is a production of aptly named *second messengers* within the cell. The best known of these is cyclic 3′, 5′ -adenosine monophosphate (cyclic AMP).[3] It appears that the result of a hormone-receptor interaction is the stimulation of the enzyme adenylate cyclase, which occurs in the membrane, presumably very close to the receptor itself. This enzyme is accessible to the components of the cytoplasm and, when activated, it catalyzes the synthesis of cyclic AMP from ATP. The structures of these compounds are shown in figure 7.7. It is thought that the activation process which occurs when a hormone interacts with its receptor is not quite direct. There is increasing evidence that activation involves an additional component in the membrane interposed between the receptor and adenylate cyclase, and known as the *transducer*. The transducer's activation of adenylate cyclase may not be an all-or-nothing process but can be modulated by the presence of other substances.

Also occurring within cells is another cyclic nucleotide, cyclic 3′, 5′ -guanosine monophosphate (cyclic GMP); its structure is shown in figure 7.7. The level of cyclic GMP also changes in response to hormonal stimulation in many systems. So far, its role in hormonal action is not as clear as that of cyclic AMP, but it may exert feedback effects on processes regulating the intracellular level of calcium.

Calcium ions are also of great importance in regulating a wide range of cellular activities. The concentration of these ions within the cell is affected directly or indirectly in virtually all cases of hormonal action. We thus have internal controlling agents in cells of two main types, cyclic nucleotides and calcium. These agents do not operate in isolation from one another, but instead interact in a cooperative manner to regulate activity.[4]

The salivary glands of the blowfly to which we referred earlier make an ideal model system for studying the interactions of cyclic AMP and calcium, as the salivary glands cells are stimulated to secrete fluid by 5-HT (which is either the hormone actually used *in vivo* or is a close mimic of it). The mechanism of fluid secretion has been worked out in some detail;[5] the currently accepted model is shown in figure 7.8. The main driving force is

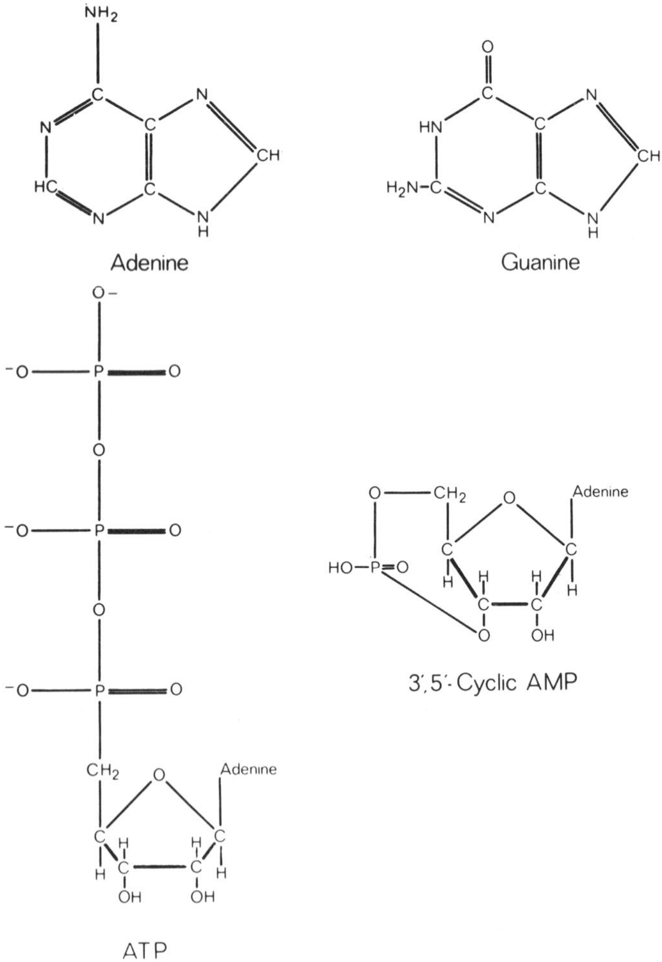

Figure 7.7 The structures of adenine, guanine, ATP, and cyclic AMP. The structure of cyclic GMP is obtained by substituting guanine for adenine in cyclic AMP.

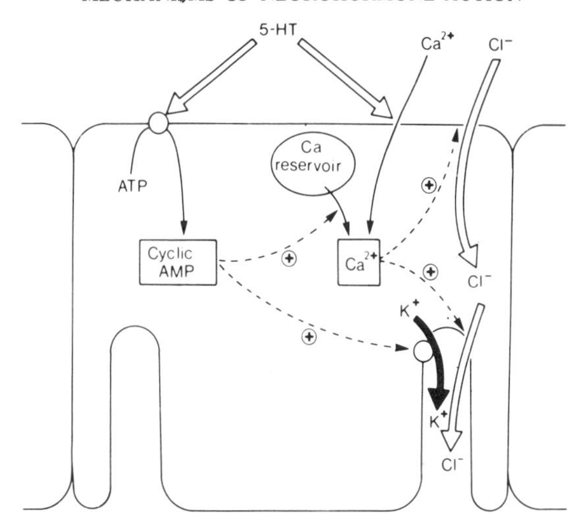

Figure 7.8 The intracellular events thought to underlie 5-HT-stimulated fluid secretion by salivary glands of *Calliphora* (after Berridge, 1975, ref. 4).

provided by an active transport of potassium ions across the cell wall into the lumen of the gland. Chloride ions passively follow the potassium ions, and the movement of these two ions causes water flow in the same direction by a mechanism which is at the moment poorly understood, but which may involve osmosis and/or electro-osmosis. Since cyclic AMP stimulates potassium transport, while calcium ions regulate chloride permeability, and since changes in potassium transport and chloride permeability cause changes in membrane potential, it is possible by using intracellular microelectrodes to follow continuously the intracellular action of the two second messengers.

Several pieces of evidence show that cyclic AMP is involved in the action of 5-HT. Not only can exogenous cyclic AMP (and some of its structurally related derivatives) stimulate fluid secretion, but during the action of 5-HT, the intracellular level of cyclic AMP increases by a factor of 2–3 times. The drug theophylline, which inhibits the intracellular enzyme phosphodiesterase responsible for the breakdown of cyclic AMP, can by itself stimulate fluid secretion. It presumably acts by allowing the accumulation of cyclic AMP produced by unstimulated slow synthesis.

Although cyclic AMP applied to the glands can stimulate fluid secretion, the potential changes recorded differ from those seen during 5-HT stimulation. The trans-epithelial potential difference is much more positive than after 5-HT treatment. The explanation turns out to be that the permeability of the cell membranes to chloride ions remains relatively low during stimulation with cyclic AMP. Further studies showed that 5-HT stimulates increases in chloride permeability by using calcium as a second messenger.

The interactions of the two second messengers first became clear in experiments where salivary glands were stimulated in calcium-free media. Under these conditions, fluid secretion is apparently normal for a few minutes, but then it steadily fails. Normal secretion is resumed as soon as calcium ions are added to the bathing medium. The reason why the glands are initially able to maintain secretion in a calcium-free medium is, probably, that they make use of calcium ions from an internal store. Once this store is depleted, the glands can only function in calcium-containing solutions. Mobilization of calcium from intracellular sites seems to occur only during stimulation and not in resting glands. Since both 5-HT *and* cyclic AMP could increase the efflux of ^{45}Ca from glands pretreated with this isotope, it appears that an important action of cyclic AMP is to release calcium from intracellular reservoirs. The difference between 5-HT action, and that of cyclic AMP alone, may be that 5-HT in addition to stimulating cyclic AMP synthesis within the cell may also cause an influx of calcium from the external medium. The intracellular interaction between cyclic AMP and calcium is shown in figure 7.8.

Although not as fully worked out as in this insect system, there is evidence that vertebrate neurohormones also depend for their action on cooperative effects of intracellular cyclic AMP and calcium ions. Perhaps the best example is the response of the toad bladder to vasopressin. Such treatment stimulates it to take up sodium (chloride) and water, and the response is dependent on the action of both cyclic AMP and calcium.

Before leaving this topic, it is worth pointing out that some of the advantages of the use of second messengers within the cell are shown particularly clearly in transporting epithelia.[6] In these systems, hormonal stimulation leads to near simultaneous changes at both sides of the cells. Not only can this readily be achieved by the use of intracellular messengers, but it is also of significance in that accelerated transport is thereby less likely to affect the intracellular environment. There is often a tendency in discussing mechanisms of epithelial transport to forget the effects that rapid transport might have on the cells responsible. If the intracellular milieu is not to be dramatically changed during hormonal stimulation, effects at, say, the apical surface need to be matched with others at the basal surface. Thus one would expect hormonal effects to involve changes at both cell surfaces. Indeed this requirement may have been of importance in favouring the evolution of intracellular messengers able to affect all parts of the cell more or less simultaneously in response to the arrival at the cells of a hormone which itself is unable, within a sufficiently short time, to reach and affect more than one part of the cell surface.

An excellent example of the way in which hormonal stimulation can lead to greatly increased transport through the cells (while preserving their internal composition) comes again from work on the blowfly salivary gland. In experiments with intracellular K^+-selective microelectrodes, it has been possible to follow changes in the cellular content of potassium ions.[7] In a bathing solution containing abnormally low potassium concentrations, unstimulated glands become depleted of potassium. On treatment with 5-HT, however, there is a rapid increase in the intracellular potassium activity (figure 7.9). One previously unsuspected effect of 5-HT is therefore a stimulation of potassium uptake across the surface facing the bathing solution. The cell potassium activity increases in spite of the fact that potassium transport out of the cell into the lumen of the gland is also very much stimulated. In glands stimulated while bathed in media containing more normal levels of potassium, the intracellular potassium activity of the glands scarcely changes at all, in spite of the very large changes in the rate at which potassium-rich fluid is secreted (figure 7.9). This neatly shows how very well the potassium fluxes across the basal and apical cell walls are

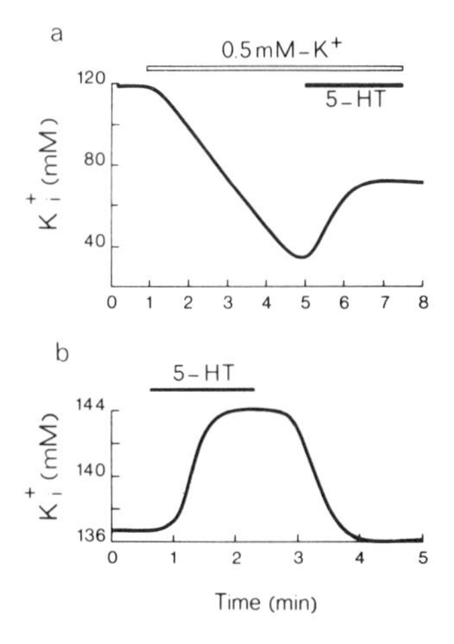

Figure 7.9 The intracellular activity of potassium ions in cells of the salivary gland of *Calliphora* during 5-HT stimulation. In (*a*), the gland is exposed to a K-deficient medium before and during stimulation. In (*b*), the bathing medium contains a normal level of potassium; note that the potassium changes by less than 8 mM (less than 6%) during stimulation (after Berridge & Schlue, 1978, ref. 7).

normally balanced, and makes the point again that hormonal stimulation leads to coordinated changes at different cell surfaces.

These last examples make very clear the advantage of a system in which the response to a hormonal signal is by a series of well-coordinated internal changes regulated by intracellular agents whose concentration is responsive through the cell membrane to the controlling hormone.

REFERENCES

1. Maddrell, S. H. P., Pilcher, D. E. M. & Gardiner, B. O. C. (1971) "Pharmacology of the Malpighian tubules of *Rhodnius* and *Carausius*: the structure-activity relationship of tryptamine analogues and the role of cyclic AMP," *J. exp. Biol.* **54,** 779–804.
2. Berridge, M. J. & Prince, W. T. (1974) "The nature of the binding between LSD and a 5-HT receptor: a possible explanation for hallucinogenic activity," *Br. J. Pharmac.* **51,** 269–278.
3. Pastan, I. (1972) "Cyclic AMP," *Sci. Am.* **227,** 97–105.
4. Berridge, M. J. (1975) "The interaction of cyclic nucleotides and calcium in the control of cellular activity," *Adv. in Cyclic Nucl. Res.* **6,** 1–98.
5. Berridge, M. J., Lindley, B. D. & Prince, W. T. (1976) "Studies on the mechanism of fluid secretion by isolated salivary glands of *Calliphora,*" *J. exp. Biol.* **64,** 311–322.
6. Maddrell, S. H. P. (1977) "Hormonal action in the control of fluid and salt transporting epithelia." In *Water Relations in Membrane Transport in the Plants and Animals* (ed. A. M. Jungreis), New York & London, Academic Press.
7. Berridge, M. J. & Schlue, W. R. (1978) "Ion-selective electrode studies on the effect of 5-hydroxytryptamine on the intracellular levels of potassium in an insect salivary gland," *J. exp. Biol.* **72,** 203–216.

General references for further reading

Case, R. M. (1978) "Synthesis, intracellular transport and discharge of exportable proteins in the pancreatic acinar cell and other cells." *Biol. Rev.* **53,** 211–354. (pp. 275–310 deal with stimulus-secretion coupling).

McGuire, R. F. and Barber, R. (1976) "Hormone receptor mobility and catecholamine binding in membranes. A theoretical model." *J. Supramolecular Structure*, **4,** 259–269.

Rasmussen, H. and Goodman, D. B. P. (1977) "Relationships between calcium and cyclic nucleotides in cell activation." *Physiol. Rev.* **57,** 421–509.

CHAPTER EIGHT

FEEDBACK CONTROL OF NEUROHORMONES

Feedback mechanisms

If the action of a liberated neurohormone is to be regulated, then there must be some mechanism by which it can directly, or through its effects, influence the rate at which it is released. The need for such regulation is the more obvious when it is recalled that some neurohormones have effects which depend on their concentration in circulation (see p. 125 for example). The possible ways in which feedback effects can regulate the release of a neurohormone are shown in figure 8.1.

In chapter 3 we saw how both vertebrate peptide hormones and a peptide from the animal itself could produce long-lasting changes in the behaviour of neurosecretory cells in the brain of the snail *Otala*. It is a distinct possibility, therefore, that neurosecretory cells might be sensitive to their own hormones, and so their behaviour might *directly* be modulated by the concentration of hormone in circulation.

If such direct sensitivity occurs in animals with a blood-brain barrier, it presumably must operate at the axon endings which, as we have seen, are the only accessible parts of the neurosecretory cells in such animals. In insects there is some evidence that the neurosecretory endings are directly stimulated by blood-borne humoral factors. The clearest case is that of release of a hormone controlling egg development in adult females of the mosquito *Aedes sollicitans*.[1] The hormone is produced by median neurosecretory cells in the brain, is transported along their axons, and accumulates in the axon endings in the corpora cardiaca, the neurohaemal organs for the neurosecretory cells of the brain. When the female insect takes a blood meal, the egg development hormone is released into circulation. However, even if all nervous connections between the corpora cardiaca and the brain are cut, the hormone is still released in response to a blood meal. Even corpora cardiaca implanted into animals which have been operated upon to remove the median neurosecretory cells, can release hormone in response to a blood meal of the host.[1] This is clear evidence

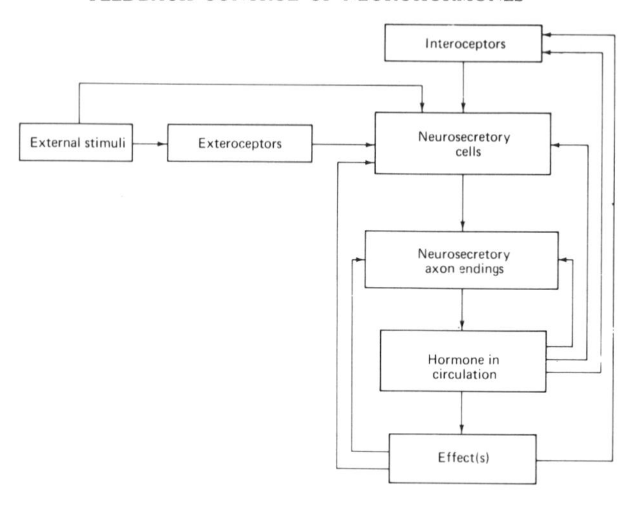

Figure 8.1 The pathways which may be involved in the control of the release of a neurohormone.

that the stimulus is humoral. The corpora cardiaca of *A. sollicitans* do not contain the intrinsic neurosecretory cells that they often do in other insects (p. 90). They do, however, have many ganglion cells which are apparently ordinary neurones and which appear to make synapses with the neurosecretory axons. So the humoral factor which provokes the release of the egg-development hormone could act either by affecting the neurosecretory terminals directly, or by stimulating the other neurones which would then control the axon endings *via* the synapses they make with them. It seems very likely that the former explanation is the correct one, because the hormone is released after feeding, even in mosquitoes from which the corpora cardiaca were removed a few days previously. As we saw earlier (p. 91), replacement corpora cardiaca are regenerated at the cut ends of the nerves leaving the brain, but they lack elements intrinsic to normal corpora cardiaca.

The evidence then suggests that the behaviour of at least some neurosecretory axon endings can be affected by a factor in circulation in the blood. We should now look at some cases where feedback control of hormone release occurs, to see how the control is effected.

In nereid worms (polychaete annelids, see p. 84) hormonal control of sexual maturity (measured as the state of development of the oocytes) involves a steady decline in the concentration of an inhibitory hormone from the brain (p. 125). The same hormone is probably involved in promoting growth by the formation of new segments, and in stimulating regeneration of segments after damage. If brains from worms of different age are implanted into other worms, also of different age, the results show that the release of the inhibitory hormone declines with age. Experiments involving direct measurements of the amounts of hormone extractable from the brain,[2] and culturing brains in isolated parapodia with developing gonads,[3] also showed that less hormone is produced in older brains. Although, of course, the decline in hormone production probably involves some autonomic control, there is evidence of negative feedback from the developing gonads. The evidence is as follows.

The brains from mature worms are normally inactive in promoting regeneration and inhibiting gonadal development. However, if they are transplanted into young host worms, they become activated and, if they are moved through a series of very immature worms, they will release hormone for greatly extended periods. If mature ovaries are implanted into immature hosts, the gonads of the young worms rapidly develop and the worms lose their ability to regenerate lost segments. The more mature is the implanted tissue, the more effectively is the secretion of brain hormone reduced. Mimicking the presence of mature eggs by implanting small glass spheres did not reduce the release of brain hormone. It seemed, therefore, that humoral effects of the maturing eggs, rather than the mechanical distension they produced, were responsible for their effect on the brain. Further details of the mechanism of feedback are not available, but it has been speculated that feedback is through cells in the infracerebral gland (p. 85), which lies in intimate contact with the coelomic fluid.[4]

In the insect *Rhodnius prolixus* there have been studies of the control mechanisms involved in the neurosecretory regulation of growth and moulting. The median neurosecretory cells of the brain produce a hormone, the prothoracicotrophic hormone (PTTH), which is released from the corpora cardiaca. This hormone stimulates the prothoracic glands to synthesize and release the moulting hormone ecdysone.* In the fifth instar larva, release of brain hormone is necessary for a period of eight days after the blood meal if ecdysis is to occur. After this critical period, the continued release of PTTH is unnecessary, and at this time there are changes in the appearance of the median neurosecretory cells suggesting that synthesis of PTTH had been stimulated and its release inhibited. Since the principal known effect of PTTH is to stimulate the production of moulting hormone,

* It is possible that, as in other insects, it is α-ecdysone which is released and special cells (oenocytes) present elsewhere in the body convert this to the much more active β-ecdysone.

it could well be that moulting hormone might have feedback effects on the cells releasing PTTH. This idea was tested in two ways. In the first, an insect preparing to release PTTH was joined in parabiosis (an arrangement whereby the two blood spaces are made confluent) with an insect in which release had occurred for the critical period, and the haemolymph in consequence contained moulting hormone. The median neurosecretory cells of the younger insect were induced to switch from their normal sequence of cytological events to one characteristic of an insect at a much later stage.[5] Very similar results were obtained from experiments in which

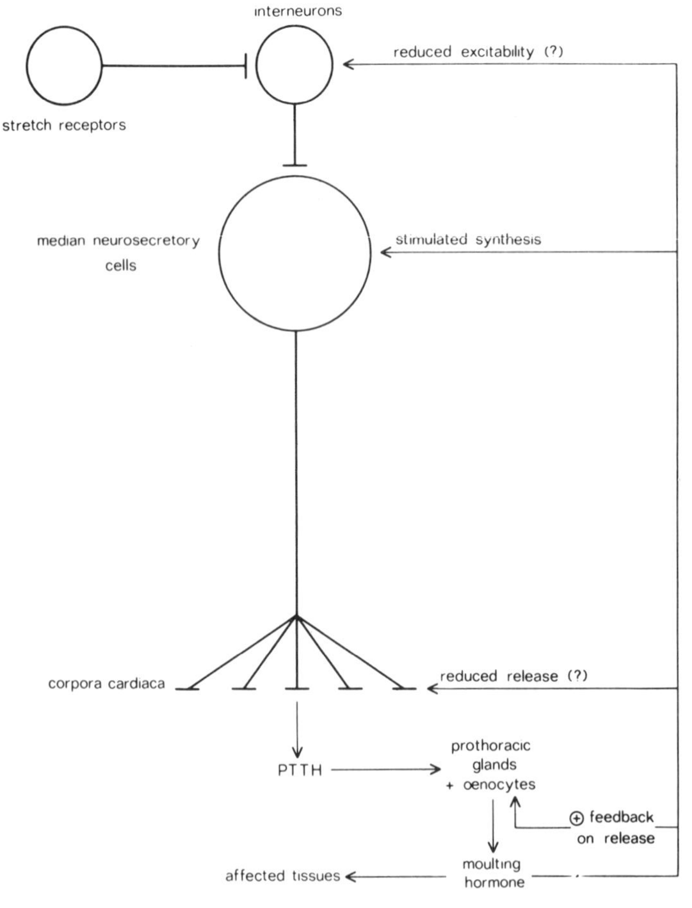

Figure 8.2 Possible feedback effects in the control of the release of PTTH and moulting hormone in *Rhodnius* (after Steel, 1975, ref 6).

recently fed insects were injected with moulting hormone.[6] The simplest hypothesis is that moulting hormone has inhibitory effects on PTTH release, but it could be that changes in the concentration of some other substances in the haemolymph might be responsible. It will be of interest if it should turn out that moulting hormone acts directly as the agent of negative feedback, as there is also evidence that it seems to have a positive feedback on its own production, thereby amplifying the hormonal instructions from the brain. A scheme to show the proposed interactions between the median neurosecretory cells and the moulting hormone is shown in figure 8.2.

Although not strictly a case of feedback control, it is worth mentioning here another instance of the regulation of the activity of neurosecretory cells by factors in the haemolymph. The activity of the median neurosecretory cells of the brain of the adult blowfly *Calliphora erythrocephala* is much affected by the presence in the haemolymph of the juvenile hormone released from glands known as the corpora allata.[7] Removing these glands reduced the activity, and implantation of active corpora allata stimulated the secretory activity, of the median neurosecretory cells. A hormone from these neurosecretory cells and juvenile hormone are together involved in controlling egg development. The effects of the juvenile hormone in stimulating release of the other hormone involved thus make good sense in coordinating the release of the two hormones jointly involved in controlling reproduction.

In the house-fly *Musca domestica* there is evidence that negative feedback from the mature eggs cuts off the release of a hormone from the brain, which controls the supply of protein to the eggs.[8] In locusts there are effects similar to those described above for *Calliphora*; removing the ovaries, starvation, or removing the corpora allata all lead to a reduced rate of release of hormone from the brain.[9] Again there is a coordination of the activities of the release sites for the two hormones important in reproduction. The cross effects presumably occur in the same way that feedback effects are achieved.

Finally, it is worth mentioning some evidence for a positive feedback process in the locust *Anacridium aegyptium*.[10] A short period of electrical stimulation of the median neurosecretory cells is enough both to start and maintain vitellogenesis and egg maturation. It seems that a positive feedback loop may be established, leading to a sustained release of neurosecretion from the brain for as long as is necessary to complete the maturation of the batch of eggs.

What these invertebrate examples show, then, is the following:

Neurosecretory cells can be directly affected by humoral agents in the blood. These effects may be mediated at or close to the neurosecretory axon endings. Both release and synthesis of neurosecretion may be affected.

The evidence is that the feedback effects on the release of a particular neurohormone all involve blood-borne factors other than the neurohormone itself. In other words, the mechanism is an indirect one acting through the effects of the hormone, rather than a direct action of the liberated hormone on its own release sites. From the point of view that the purpose of hormone release is to produce an effect, of course, feedback control through the effect makes good sense.

How do feedback effects in vertebrates compare with these invertebrates cases? We have seen (p. 37) how in the rat, release of vasopressin is induced by increases in the osmotic concentration of the blood plasma. When the osmotic concentration falls again, the vasopressin release rate gradually returns to its normal low value. Plainly here the feedback response acts from an effect of the hormone release. In this case it is supposed that the effect is perceived by osmoreceptors, long thought to be neurones in the supraoptic nuclei of the brain. However such neurones have not yet been found in the hypothalamus.

It is with the hypothalamic regulatory peptides (hypophysiotrophic hormones, see p. 134) that more complex and interesting feedback effects occur.[11] These neurohormones regulate the release of such hormones as ACTH, growth hormone, prolactin, etc., from the anterior pituary (adenohypophysis). They are released from a group of neurosecretory cells (the parvocellular system) in the basal medial part of the hypothalamus.

The events concerned in the release of corticotropin-release factor (CRF) are an excellent illustration of the way in which the system is regulated. CRF released in the basal part of the hypothalamus is carried to the adenohypophysis in the capillaries of the hypothalamic-hypophysial portal system. There the cells which release adrenocorticotropin (ACTH) are stimulated and ACTH appears in the general circulation. As a result, the adrenal cortex releases its steroid hormones, the corticosterones. These events are illustrated in figure 8.3, as are the feedback pathways which have been proposed. There is good evidence that corticosterones in the plasma inhibit release of ACTH from the pituitary. They may also suppress CRF release by acting in the brain. Finally there is some evidence that ACTH may inhibit CRF release. As suggested earlier (p. 134, 157) it certainly seems that a control system involving more than one step is more open to feedback control!

The control of thyroid-stimulating hormone (TSH) by thyrotropin-

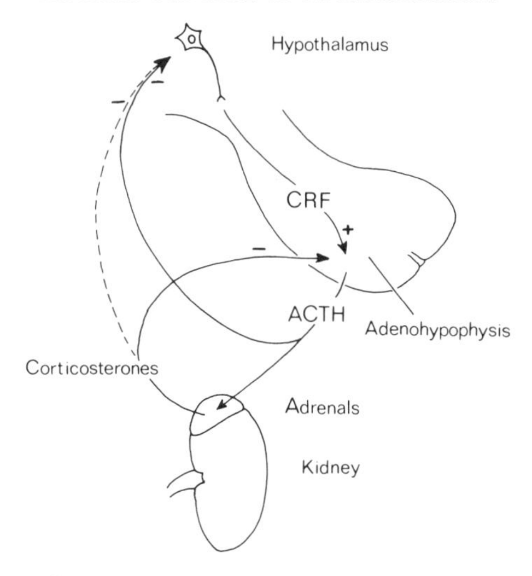

Figure 8.3 The control mechanisms regulating the level of corticosterones in circulation. Note that both stimulation (+) and inhibition (−) are involved.

releasing factor (TRH) is regulated by feedback effects of thyroxine on the release of TSH and possibly also on TRH release. This conclusion is mostly based on the results of experiments in which thyroxine levels in the plasma were increased by injecting extra thyroxine. Unusually, however, there is also some evidence that if thyroxine plasma levels are experimentally reduced, then thyroxine release is stimulated.

Prolactin release seems to be regulated by two neurosecretory factors: prolactin-releasing factor (PRF) which stimulates prolactin release, and prolactin-inhibitory factor (PIF) which inhibits the release of prolactin. Increased levels of prolactin stimulate the release of PIF. It remains to be seen whether declining levels of prolactin act to promote the release of PRF and/or inhibit the release of PIF.

Finally, it is worth mentioning that there are indications that growth hormone, whose release is stimulated by the growth hormone-releasing factor (GRF) may be under a feedback control in which it inhibits its own release. Apart from this one exception it seems that, as in invertebrates, all feedback regulation of hormone release in vertebrates is achieved other than by a direct effect of the circulating hormone.

Research into feedback control of hormone release is a very active area at the moment, and we can look forward to exciting developments in the future. In the medical field the results will be of great interest in understanding and, hopefully, treating cases of hormone imbalance or deficiency. Work on the control of neurosecretory systems in insects could well lead to the development of a range of more sophisticated and selective insecticides.[12]

REFERENCES

1. Meola, R. & Lea, A. O. (1971) "Independence of paraldehyde-fuchsin staining of the corpus cardiacum and the presence of the neurosecretory hormone required for egg development in the mosquito," *Gen. Comp Endocrinol.* **16,** 105–111.
2. Durchon, M. & Porchet, M. (1971) "Premières données quantitatives sur l'activité du cerveau des Néréidiens au cours de leur cycle sexuel," *Gen. Comp. Endocrinol.* **16,** 555–565.
3. Porchet, M. (1971) "Variation de l'activité endocrine des cerveaux en fonction de l'espèce de sexe et du cycle vital chez quelques Néréidiens (Annélides Polychètes)," *Gen. Comp. Endocrinol.* **18,** 276–283.
4. Baskin, G. (1974) "Further observations on the fine structure and development of the infracerebral complex ('Infracerebral gland') of *Nereis limnicola* (Annelida, Polychaeta)," *Cell Tissue Res.* **154,** 519–531.
5. Steel, C. G. H. (1973) "Humoral regulation of the cerebral neurosecretory system of *Rhodnius prolixus* (Stal) during growth and moulting," *J. exp. Biol.* **58,** 177–187.
6. Steel, C. G. H. (1975) "A neuroendocrine feedback mechanism in the insect moulting cycle," *Nature* **253,** 267–269.
7. Thomsen, E. & Lea, A. O. (1968) "Control of the medial neurosecretory cells by the corpus allatum in *Calliphora erythrocephala,*" *Gen. Comp. Endocrinol.* **12,** 51–57.
8. Adams, T. S., Grugel, S., Ittycheriah, P. I., Olstad, G. & Caldwell, J. M. (1975) "Interactions of the ring gland, ovaries and juvenile hormone with brain neurosecretory cells in *Musca domestica,*" *J. Insect. Physiol.* **21,** 1027–1043.
9. McCaffrey, A. R. & Highnam, K. C. (1975) "Effects of corpora allata on the activity of the cerebral neurosecretory system of *Locusta migratoria, migratorioides R. & F.,*" *Gen. Comp. Endocrinol.* **25,** 358–372.
10. Girardie, A., Moulins, M. & Girardie, J. (1974) "Rupture de la diapause ovarienne *d'Anacridium aegyptium* par stimulation éléctrique des cellules neurosécrétrices médianes de la pars intercerebralis," *J. Insect Physiol.* **20,** 2261–2275.
11. Yates, F. E., Russell, S. M. & Maran, J. W. (1971) "Brain-adenohypophysial communication in mammals, *A. Rev. Physiol.* **33,** 393–444.
12. Maddrell, S. H. P. & Reynolds, S. E. (1972) "Release of hormones in insects after poisoning with insecticides," *Nature*, **236**, 404–406.

CHAPTER NINE

SUMMARY AND CONCLUSIONS

THIS CONCLUDING CHAPTER HAS TWO MAIN OBJECTIVES: TO SUMMARIZE WHAT is known of each aspect of neurosecretory systems and to pick out what seem to us to be the most urgent problems for future research.

Neurosecretory hormones are first synthesized in the cell body as polypeptides of high molecular weight, and high concentrations of these are packed into membrane-bound granules. These granules are moved down to the axon endings by an axonal transport process. Axons convey materials along their length at different speeds, but for neurosecretory granules is reserved first-class travel by the fastest transport mechanism. It seems that this involves an association with the microtubules which run along axons. Details of this mechanism are not yet known.

The most recently made granules move first into the axon endings, the sites of hormone release. If they are not released fairly soon, they then move away into other areas of the axon which are often dilated and where it is presumed the ageing granules are broken down. How this traffic is regulated is not known, nor is the fate of the products of granule breakdown.

When hormones are released at high rates there follows an increase in the rate of synthesis of new granules in the neurosecretory cell bodies. Nothing is known of how the rate of granule synthesis is controlled.

During the movement of neurosecretory granules from the cell bodies to the axon endings, the polypeptide in them is cleaved into the active hormone and one or more other peptides or proteins. In the neurohypophysis, the proteins are known as *neurophysins*. Neurophysins are thought to aggregate with their hormones so as to decrease the risk of the hormones (which are often small peptides) being lost by diffusion out of the granule; the aggregation may also reduce the total osmotic activity of the granule contents and so prevent osmotic swelling. Mature granules often contain small peptides other than the active hormone. The function of these extra peptides is unknown; one suggestion is that they may modulate the mechanism of release of the granule contents.

Because most neurosecretory cells release their active products into enormously larger volumes than do conventional neurones, they have a series of adaptions to enable them to release sufficient hormone to have an effect. The number of molecules of hormone packed into a granule is close to the theoretical maximum, and large numbers of granules are available in the swollen axon endings from which release occurs. The rate of release from one ending seems to be a limiting factor; however, it is overcome by a great increase in the number of endings that each cell has. In some cases the number of endings per cell is thought to be as high as two thousand. Finally, the hormones themselves are often very active compounds, capable of having effects at concentrations as low as 10^{-12}—10^{-13}M.

Neurosecretory cells like other nerve cells are electrically active. They conduct impulses along their axons, and it is the resulting depolarization of the cell membrane at the axon endings which initiates hormone release. In some cases the rate of hormone release is now known to depend on the intensity of impulse traffic in the axon, the most effective firing pattern being one in which the spikes arrive in bursts or volleys. There is, however, scope for more research on the exact relation between firing rate and rate of hormone release. Action potentials recorded from the cell bodies and axon terminals are considerably longer than in conventional neurones. The initial inward current at these points in the cell is carried by both calcium and sodium ions. By contrast the axonal potentials are shorter and are thought to require inward movement only of sodium ions.

The firing of neurosecretory cells is affected both by synaptic input and by humoral agents. The humoral effects suggest that research into possible feedback regulation of neurosecretory cells would be rewarding. In addition more work is needed on how the diurnal rhythm in the firing pattern of some neurosecretory cells is maintained and controlled.

The arrival of an impulse at a neurosecretory axon ending depolarizes it and causes an influx of calcium ions. The rise in internal calcium concentration triggers granule release by exocytosis, a process in which the granule membrane fuses with that of the axon, so that the granule's contents are released into the extracellular space. There is both ultrastructural and biochemical evidence for exocytosis. What is not yet clear is how an increase in calcium ion concentration causes exocytosis. It seems likely that a decrease in cytoplasmic viscosity is one element in the process; it may allow granules to approach the axon surface more easily. The events occurring during membrane fusion particularly need clarification.

After exocytosis the extra membrane incorporated into the axon surface has to be retrieved. In many cases this seems to be achieved by the formation of electron-lucent vacuoles similar in size to those of the neurosecretory granules before they have released their contents. However, other evidence suggests a retrieval mechanism involving the production of much smaller vesicles. Plainly more research is needed here. Apart from the fact that depolarization of the axon endings is not a sufficient stimulus, nothing is known of how the retrieval process is started.

Neurosecretory cells in different animals all function in a basically similar way along the lines summarized in the above paragraphs. Where they differ in the different animals is in their morphology, and in their organization at the supracellular level. Broadly speaking, smaller animals have fewer neurosecretory cells, each of which has fewer endings and the hormones released are less active. This is likely to be due to the inescapable fact that the surface area to volume ratio of any structure in an animal decreases as size increases. It is thus "easier" for a small animal to release effective quantities of hormone into its extracellular space than it is for a larger animal. In large animals, neurosecretory cells tend to have very large numbers of axon endings (release sites); as many as 2000 per cell are found in the vertebrate neurohypophysis. An exception to this occurs in the adrenal system, where the neurosecretory cells have no axon and so no endings. In this case an effective release area is achieved by a vast increase in the number of secretory cell bodies.

In animals with reduced extracellular space and no circulatory system, neurosecretory hormones can be used to set up gradients of concentration. Such gradients can be used by responding cells to learn of their position within the organism. Not surprisingly, in animals with only a poor circulation, neurosecretory hormones are mostly used to regulate long-term changes rather than the more rapidly changing metabolic processes.

Most neurosecretory cells lie in the central nervous system. Where there is a "blood-brain barrier", the cells are not easily able to release their products into circulation. Rather than breach the barriers round the ganglia, the neurosecretory axons tend to gather together and run some distance away from the ganglia down the nerve cords or in the peripheral nerves, before they penetrate the blood-brain barrier. Having the breach at a distance presumably makes it easier to maintain a suitably constant and controlled environment for the proper functioning of the nerve cell bodies that lie in the ganglia. Once the neurosecretory axons have together crossed the barrier, they branch profusely and are separated from the extracellular fluid only by highly permeable connective tissue which serves only to support and maintain the structural integrity of the resulting neurohaemal organ. In animals where there is no blood-brain barrier, neurosecretory axon endings are usually much more diffusely arrayed, often, for example, being dotted over the surface of the ganglion in which the cell bodies lie.

Although most neurosecretory cells lie in central nervous systems, they may also occur in or on peripheral nerves. Such cells in invertebrates are not homologous with those of vertebrates (the APUD series). Octopamine-secreting cells in the lobster have release sites at two distinct positions along their axons, an arrangement whose significance is not known.

In invertebrates, neurohormones act to regulate a very wide range of functions. The use that vertebrates make of neurohormones is much more limited. One can speculate that part of the reason for this is the generally larger size of vertebrates. Even with more neurosecretory cells, each with large numbers of endings, it may not be easy for a large animal rapidly to release an effective concentration of hormone. It may, in evolution, have proved more convenient and involve less expenditure to use conventional direct nervous control for the regulation of most systems requiring adjustment on a relatively short-term basis.

Neurohormones can be used to convey more information than one might at first suppose. Many target systems respond in qualitatively different ways to different concentrations of hormone and to different lengths of exposure to a hormone. A single hormone may act to control several different organs or tissues which may have separate roles in a single overall physiological process.

This is not the place to list again the roles of all the different neurohormones. However, for two classes of neurohormones, in particular, details of roles and status are missing or incomplete. These are the peptides, known (or thought) to be released by the pineal organ and urophysis of vertebrates. More research is needed here.

The evidence is that neurohormones act by transient binding to specific receptors on the surfaces of cells of the target organs. The short-lived nature of the association ensures that the response of the cells can readily be regulated by changes in hormone concentration, and it means also that the response can be terminated rapidly as hormone is removed from the extracellular fluid. Neurohormone-receptor interaction at the cell surface leads to concomitant changes in internal, so-called *second messengers*. This allows all parts of the cell to respond in an integrated way to an external stimulus. The internal agents involved are those almost universally found to be involved in intracellular regulation—calcium ions and the cyclic nucleotides, cyclic AMP and cyclic GMP.

Index

Index